Estates Gazette

Planning Law Reports

2005

Volume 1

Edited by
Barry Denyer-Green LLM PhD FRICS
of Middle Temple, barrister
and
Navjit Ubhi LLB (Hons)
of Middle Temple, barrister

Hélène Préchac
Publisher's editor

2005

Routledge
Taylor & Francis Group
LONDON AND NEW YORK

First published 2005 by Estates Gazette

Published 2014 by Routledge
4 Park Square, Milton Park, Abingdon, Oxon OX14 4RN
605 Third Avenue, New York, NY 10017

Routledge is an imprint of Taylor & Francis Group, an informa business

© Taylor & Francis,
Barry Denyer-Green and Navjit Ubhi, 2005

All rights reserved. No part of this book may be reprinted or reproduced or utilised in any form or by any electronic, mechanical, or other means, now known or hereafter invented, including photocopying and recording, or in any information storage or retrieval system, without permission in writing from the publishers.

Notices
Knowledge and best practice in this field are constantly changing. As new research and experience broaden our understanding, changes in
research methods, professional practices, or medical treatment may become necessary.

To the fullest extent of the law, neither the Publisher nor the authors, contributors, or editors, assume any liability for any injury and/or damage to persons or property as a matter of products liability, negligence or otherwise, or from any use or operation of any methods, products, instructions, or ideas contained in the material herein.

Product or corporate names may be trademarks or registered trademarks and are used only for identification and explanation without intent to infringe.

ISBN 13: 978-0-7282-0469-0 (pbk)

ISSN 0954-1977

CONTENTS

CASES REPORTED IN THIS VOLUME

INDEX OF SUBJECT MATTER

CASES JUDICIALLY CONSIDERED

For case references see pages shown in bold type

R (on the application of Cobbledick) *v* First Secretary of State and another

[2004] EWHC 1341 (Admin)

ADMINISTRATIVE COURT
Sullivan J
17 May 2004

Costs – Enforcement notices – Appeal allowed – Application for costs A **– PPG 18 – Whether local planning authority issuing enforcement notice unreasonably – Whether Secretary of State unreasonably dismissing application for costs – Whether council acting unreasonably in failing to make preliminary enquiries**

The claimant appealed against an enforcement notice issued by the second B defendant local planning authority. That appeal was allowed by the inspector on the ground that planning control had not been breached. The local planning authority's appeal against that decision was dismissed by Sullivan J (*North Cornwall District Council* v *Secretary of State for Transport, Local Government and the Regions* [2002] EWHC 2318 (Admin); [2003] 1 PLR 28). The claimant's application to the Secretary of State, asserting that C the local planning authority should pay the costs of his appeal before the inspector, was dismissed. The claimant challenged that decision contending, *inter alia,* that the council's failure to make any enquiry prior to embarking upon enforcement action was unreasonable in the light of PPG 18 Enforcing Planning Control.

Held The application was allowed. D

The Secretary of State's decision letter had failed to address the essential points. Had the local planning authority made the most elementary enquiries of the claimant, service of the enforcement notice, against which he successfully appealed, would have defied common sense. The authority had acted unreasonably in issuing an enforcement notice in the terms that it did. E

Case referred to in the judgment

North Cornwall District Council v *Secretary of State for Transport, Local Government and the Regions* [2002] EWHC 2318 (Admin); [2003] 1 P&CR 25; [2003] 1 PLR 28; [2003] JPL 600

F

Judicial review

This was an application for judicial review by the claimant challenging the First Secretary of State's refusal to award costs upon the claimant's successful appeal against an enforcement notice issued by North Cornwall District Council.

James Maurici (instructed by Bond Pearce, of Plymouth) appeared for the claimant, Peter Cobbledick.

A *James Strachan (instructed by the Treasury Solicitor) represented the first defendant, the First Secretary of State.*

North Cornwall District Council, an interested party, did not appear and were not represented.

The following judgment was delivered.

B **SULLIVAN J:**

[1] This application for judicial review is a sequel to *North Cornwall District Council* v *Secretary of State for Transport, Local Government and the Regions* [2003] EWHC 2318 (Admin); [2003] 1 P&CR 25[1]. In that case, I dismissed the council's appeal against a decision of an inspector appointed by the Secretary of State allowing an appeal by Mr Peter Cobbledick
C against an enforcement notice that had been issued by the council on 1 August 2001. I described the council's appeal as a veritable storm in a very small planning teacup.

[2] Now, another storm has blown up in the teacup over the Secretary of State's decision on Mr Cobbledick's application for the costs of his successful appeal against the enforcement notice. On 25 February
D 2003, the Secretary of State exceptionally issued a "minded" decision, explaining why he proposed to make a full award of costs against the council.

[3] Having received further representations from the parties, the Secretary of State issued his final decision in a letter dated 4 July 2003. In that letter, he concluded that no award of costs against the council on
E the ground of unreasonable behaviour resulting in wasted or unnecessary expense was justified. He therefore refused Mr Cobbledick's application for costs, and, in this application for judicial review, Mr Cobbledick challenges that decision.

[4] The appeal against the enforcement notice was dealt with by written representations, so Mr Cobbledick's costs could not have been very large.
F In *R (on the application of Ealing London Borough Council)* v *Secretary of State for the Environment* [1999] EWHC 345 (Admin) [sic], I said that where the costs of judicial review proceedings are likely to be equal to, if not more than, the costs the subject of the award, the parties should think long and hard before deciding to seek permission for judicial review. That advice was particularly directed at local planning authorities that sometimes
G seem to view an adverse award of costs as a slur on the council's corporate reputation as a planning authority that have to be expunged regardless of the sums at stake. In that case, the award was made by an inspector who had conducted the enquiry and who was therefore particularly well placed to decide whether the council, in giving their evidence, had behaved unreasonably.

H [5] In the present case, because the inspector had allowed Mr Cobbledick's appeal on ground (c), he did not consider the planning merits of the appeal, and, following the council's unsuccessful challenge in

[1] [2003] 1 PLR 28

the High Court, Mr Cobbledick's application for costs was determined by A
the inspectorate on the basis of written representations from the parties.

[6] Although the sums at stake are likely to be small, this is no ordinary costs decision. The Secretary of State took the exceptional step of issuing a "minded" letter. That letter is very detailed and runs to no less than 26 paragraphs. The final decision letter is similarly detailed and runs to 24 paragraphs. In these circumstances, it behoves the Secretary of State B
to get the right answer, and I do not therefore accept Mr James Strachan's invitation on behalf of the Secretary of State to refuse to grant Mr Cobbledick relief as a matter of discretion.

[7] I can take most of the background facts from the judgment in the earlier case. Mr Cobbledick is the managing director of Oceancrafts Ltd, and for the past 20 years (as at 2002), Oceancrafts had taken premises in C
the retail area of Bude to sell shells and other "ocean crafts" to visitors during the summer season. In 2001, Mr Cobbledick was unable to find a shop that would be available for the summer, so he took a licence of 14 Belle Vue, Bude (the property).

[8] The property is a two-storey, double-fronted Edwardian house in the primary commercial area of Bude. It is one of a terrace of former D
houses, 4-18 Belle Vue, all of which, apart from the property, are now in primarily retail use with single-storey extensions on the frontage, extending up to the back edge of the footway. These extensions provide additional retail floor space and incorporate modern shop fronts. The property is the exception: it has a front garden laid to grass with a footpath to a central front door flanked by a matching pair of two-storey splayed E
bay windows.

[9] Previous occupiers of the property were a firm of chartered accountants and agents for the Woolwich Building Society. They carried on business offering financial and professional services to visiting members of the public for some 36 years. It was not in dispute that their use of the property fell within Class A2 (financial and professional services) in the F
Schedule to the Town and Country Planning (Use Classes) Order 1987.

[10] On 12 July 2001, Mr Cobbledick was granted a licence to occupy the property until 5 October 2001. The property was in a poor condition, the previous occupiers having received a dilapidations notice from the owner. A substantial amount of work needed to be done to it, and the peak summer season was rapidly approaching, with the schools due to break up G
on 19 July. So, Mr Cobbledick erected in the forecourt a temporary timber structure, described by the inspector as a standard exhibition display unit, in order to allow trading to start while work was done to the main building. It took three days to erect and fit out the unit and another two days to stock it. Oceancrafts began to trade from the property on 19 July.

[11] On 1 August, the council issued two enforcement notices. Both H
related to "land at 14 Belle Vue… as shown edged red on the attached plan". The whole of the property, and not merely the temporary structure, was shown edged red.

[12] One enforcement notice alleged an unauthorised material change of use:

> The use of the land and the timber structure erected thereon for retail purposes.

[13] It required Mr Cobbledick:

> To discontinue the use of the land and the unauthorised timber structure thereon for retail sales and for exhibiting items for sale.

[14] Seven days were allowed for compliance after the notice took effect on 8 September. That would have allowed Oceancrafts to complete the works to the main building and to trade for much of the summer season, but the enforcement notice was backed by a stop notice that required all retail use of the land and the unauthorised timber building thereon to cease on 5 August.

[15] The other enforcement notice alleged unauthorised operational development, "The erection of an unauthorised timber structure currently being used as a shop", and required its removal within seven days after the enforcement notice took effect on 8 September.

[16] Mr Cobbledick complied with the stop notice, stopped trading and removed the temporary structure. He appealed against both enforcement notices, but since the temporary structure had been removed in compliance with the second enforcement notice, the appeal against that enforcement notice was not pursued.

[17] His appeal against the first enforcement notice was dealt with by way of written representations. The inspector visited the site on 1 May 2002. In his decision letter dated 13 May 2002, the inspector allowed the appeal on ground (c), namely that there had been no breach of planning control, and did not therefore consider the planning merits under ground (a).

[18] In the "Minded" decision letter, the Secretary of State said:

> 20. The Secretary of State sees no documentary evidence that the Council, either before or after the appeal, addressed the issue, raised in your appeal statement, of whether the alleged breach of control had seriously affected public amenity. The same applies to your client's evident concern that the LPA's action should be commensurate with the alleged breach of control. He had not, it seems, received any complaint nor been told of any complaint to the Council; nor had Council officers approached him to discuss the matter before taking formal action. The Council have not rebutted these points, which are considered significant in the light of:
> - The guidance in paragraph 5(3) to (5) of PPG 1.
> - The evidence of the temporary nature of the structure causing concern.
> - The impact on your client's business of taking precipitate enforcement action linked with a stop notice.

[19] Subparagraphs (3) to (5) of para 5 in PPG 18[1] are concerned with the general approach to enforcement. They advise local planning authorities that are considering enforcement action to consider whether the breach of development control would unacceptably affect public amenity. They advise that enforcement action should always be commensurate with a

[1] Enforcing Planning Control

breach of planning control, and they also indicate that local authorities may seek to persuade owners or occupiers of sites voluntarily to remedy the harmful effects of unauthorised development. The point is made that if those negotiations fail, they should not be allowed to delay formal enforcement action.

[20] In other parts of PPG 18, advice is given in relation to enforcement action to deal with unauthorised development by small businesses or self-employed people. Paragraph 15 makes it clear that the initial aim should be to explore in discussion with the owner or operator whether the business can be allowed to continue operating acceptably on the site at its current level of activity, or perhaps less intensively. Paragraph 16 says:

> Unless it is urgently needed, formal enforcement action should not come as a "bolt from the blue" to a small business or self-employed person. It should be preceded by informal discussion about possible means of minimising harm to local amenity caused by the business activity...

[21] Mr Cobbledick's written representations for the enforcement notice appeal had included the following information:

> It was a surprise when the Stop Notice was issued especially as it covered the whole site. The structure was temporary. Had the Council consulted with us, I could have removed it. Ideally I needed until August 15th to fully furbish the two front rooms. The temporary building could have been lifted out onto a lorry in an hour.

[22] The council responded to the minded letter by explaining what members' concern had been and what information had been placed before members. The Secretary of State was told that a video presentation had been made to the council's planning and development committee, which considered enforcement action on 24 July 2001 as an urgent item. A copy of the video was enclosed. In addition, a copy of a memorandum from the council's head of development control was also enclosed, confirming that the item was presented to the committee and that:

> the amenity issue was the primary consideration in the Members' decision.
>
> As stated by the Head of Development Control, the Enforcement Notice relating to the operational development refers to Policy DVS1 of the District Local Plan in its reasons for issuing the notice. This refers to siting and design and not respecting the character and appearance of the locality, ie: these matters constitute an amenity objection. A planning application would clearly not have been allowed.
>
> You will note from the video that the building is of timber plywood construction, with a flat roof and stained a washed-blue. It fronts onto the main road through Bude (Bude's primary shopping area).
>
> I recollect that Members were concerned that because of the speed of the building's erection, its unacceptable appearance and the visual harm of the building in the peak holiday season, it was necessary to take immediate action. The commercial season in Bude is relatively short, with the core-season being the end of July, August and the beginning of September, ie: during the school holidays.

A [23] The Secretary of State was also sent a copy of a letter of complaint from the Bude-Stratton Town Council dated 27 July 2001:

> The Town Council would like to express its concerns about the development which has taken place to the front of the property known as 14 Belle Vue, Bude. The construction is, to say the least, an eyesore situated as it is in the main trading street in Bude...

B

[24] The video was sent to Mr Cobbledick's solicitor, and in a letter dated 17 April 2003, it said:

> The further representations supported by the video do not progress the principal reason by the Secretary of State for being minded to make an award of costs. Whatever view is taken as to the effect on local amenity, the Council
> C patently failed to make any enquiry of Mr Cobbledick as to the likely duration for the siting of the temporary building. Although it is not incumbent to serve a Planning Contravention Notice, in deciding whether it is expedient to take enforcement action, it is incumbent on the local planning authority to have regard to material considerations in deciding the expediency of such action. Those considerations must include the likely duration of any harm which the Council have identified from the breach of planning control. The
> D Council have not produced any evidence to show that the Committee was made aware that the temporary building would be removed after a short period following the completion of the refurbishment works and appear to have taken their decision in the belief that the temporary building would remain for the full summer season. If the Council's officers were unaware of the intention to remove the temporary building on the completion of those works, this information could have been obtained by a simple contact with
> E Mr Cobbledick. The failure to make such enquiry is, in itself, unreasonable conduct on the part of the Council.

[25] The final decision letter stated that the claimant's complaints did have some force in relation to the totality of the council's action: that is to say in relation to the service not merely of the enforcement notice that was
F the subject of the appeal, but in relation to the service of both enforcement notices, one of them being backed by a stop notice.

[26] That conclusion is readily understandable. It is extraordinary that the council chose to serve a stop notice without having made any prior attempt to contact Mr Cobbledick. He had traded in Bude each summer for 20 years, and there is no suggestion that he was evading
G enforcement action.

[27] However, the Secretary of State's approach in the final decision letter, as amplified by Mr Strachan in his skeleton argument, was to clearly distinguish between the decision to serve the stop notice for which compensation is available under separate provisions in the Town and Country Planning Act 1990 and the decision to serve the enforcement
H notice, which was the subject of the appeal. The decision letter made the point that Mr Cobbledick:

> has a separate remedy under the Act for pursuing what he sees as the serious effect on his business of the Council's action. That is distinct from seeking reimbursement of the costs of pursuing his appeal against the relevant enforcement notice through this application for an award of costs.

[28] The Secretary of State's conclusions are set out in para 21 of the A
final decision letter:

> The conclusion reached, in all the circumstances and on the balance of probability, is that the Council did not act unreasonably in issuing, in combination with the operations enforcement notice, the second enforcement notice alleging a material change of use, against which your client maintained an appeal. While you criticise the Council for not first serving B
> a planning contravention notice (PCN) on your client, it is sufficiently clear that the Council
>
> - had received a complaint from the Town Council.
> - could draw on officer advice, including video evidence of the effect on local amenity.
> - resolved to act quickly on their assessment of the planning situation, C
> serving also a related stop notice.
>
> While the Council might have done more to contact Mr Cobbledick and clarify his intentions, it appears unlikely on the evidence available that the Council would have acted differently, given their view of the planning issues. The Secretary of State is therefore not satisfied, on the balance of the evidence, that the normal pre-conditions for an award of costs, as stated in paragraph 6 of Annex 1 to the Costs Circular, have been met in this case. He D
> therefore reaches a different conclusion from that provisionally stated in the letter of 25 February 2003.

[29] Upon the basis that the decision to serve a stop notice and the consequences flowing from that notice are a separate issue, and that one should focus upon the reasonableness of issuing, in combination with the operations enforcement notice, a second enforcement notice alleging a E
material change of use in respect of the whole of no 14, I am in no doubt that, in concluding that the council's conduct was not unreasonable, the Secretary of State has lost the wood for the trees.

[30] It is plain from the council's correspondence in response to the "minded" decision letter that their avowed purpose in serving the enforcement notices was to remove the temporary building that was F
regarded as an eyesore in Bude's main trading street, and that there was a need for urgent action because of the relatively short summer season in Bude.

[31] Mr Cobbledick's principal complaint was that the council simply never made any enquiry of him as to why he had placed the temporary building in front of no 14, and for how long he proposed to keep it there. G
His contention was that the failure to make those most basic enquiries before embarking upon enforcement action was unreasonable, in the light of the guidance in PPG 18, above.

[32] Had the council made any enquiries of Mr Cobbledick, they would have been told that the temporary building had been placed there to enable no 14 to be refurbished for retail use, and that "ideally" (from H
Mr Cobbledick's point of view), the temporary building would remain there until 15 August to allow the works to be completed, whereupon it would be removed.

[33] If the council had obtained that information, there can be no doubt whatsoever that no reasonable local planning authority armed with that

A information would have dreamt of serving an enforcement notice that took
effect on 7 September and thereafter gave seven days for compliance, and
that required not merely that retail use ceased in the temporary building,
but that it ceased throughout the whole of no 14 by 14 September. I realise
that the council were constrained by section 172(3) as to the earliest date
when an enforcement notice could take effect, but the question is whether
B it was reasonable to serve this enforcement notice to secure the removal of
this temporary building. On the basis of the factual information that could
have been obtained from Mr Cobbledick, which might even have included
an earlier date for removal than his "ideal" date of the 15 August, such an
enforcement notice would have been: (a) wholly unnecessary, since it was
proposed to remove the eyesore by 15 August; (b) wholly ineffective, since
C the holiday season would effectively be at an end by 14 September; and
(c) plainly excessive since the council's concern related to the temporary
structure and did not extend to the use of no 14 for retail purposes.

[34] Although Mr Strachan submitted that if one looked at the evidence
in the round, the council would have issued an enforcement notice in
relation to the retail use of no 14, quite apart from their desire to remove
D the eyesore presented by the temporary structure, that suggestion, which
does not appear in the Secretary of State's decision letter, is simply not
borne out by the evidence.

[35] The council's written statement for the purposes of the enforcement
notice appeal dealt with the ground (a) appeal against the enforcement
notice in these terms:

E

> It was considered by the Local Planning Authority (LPA) that the structure erected on the front of 14 Belle Vue was clearly contrary to the aims of District Local Plan Policy DVS1. The development did not respect the scale, character and appearance of the immediate surroundings and the wider environment.
>
> In the LPA's view there was no realistic prospect of the structure being
> F granted consent, either conditional or unconditional, therefore an application was not invited. The LPA did not feel that cosmetic additions to the structure would significantly reduce the harmful visual impact of the structure on the locality and certainly not to the degree that the building would be seen to comply with Policy DVS1.
>
> The change of use of the premises from an office to a shop would require the insertion of display window(s). At present it does not have the
> G appearance of a shop. The change of use would require operations to the front of the building and as the building is the most formal building in the street scene and least altered careful consideration of these alterations would be required.
>
> Similar to a number of other recent applications determined by the LPA, including some in Bude, the Authority would require details of alterations
> H to the front elevation as part of any change of use application. In addition details of how the area in front of the premises would be treated ie the forecourt area would also be necessary.
>
> Class A1 is acceptable in principle provided that there is a satisfactory scheme for converting the premises. It is important to have a shop that accords with good design principles such as those set out in the Council's Design Guide.

[36] These considered reflections on the part of the council merely A confirm the material contained in their response to the minded decision letter. It is plain that the council's concern was to secure the removal of the temporary structure because it was considered to be an eyesore in Bude's main shopping street, and the council did not want such an eyesore to be there during the short summer season. Given that those were the council's objectives, it is plain that had they made even the simplest enquiry of B Mr Cobbledick, the enforcement notice that was actually served would not have been served.

[37] As I have said, it would have been unnecessary to serve the enforcement notice in respect of the eyesore if the proposal was to remove it by 15 August. The object of protecting the amenity of Bude's main shopping street for the short summer season would not C be achieved by requiring removal of the structure or the cessation of the use of no 14 for retail purposes by 14 September; and there was no basis upon which, subject to detailed discussions about the design of display windows, the council would have objected to the retail use of no 14 itself.

[38] In my judgment, the Secretary of State's final decision letter, D long though it is, fails to grapple with these essential points. Had the council made the most elementary enquiries of Mr Cobbledick, service of the enforcement notice against which he successfully appealed, would have defied common sense. The council acted unreasonably in issuing an enforcement notice in the terms set out above, and the proposition that it is unlikely that the council would have acted differently if they E had done more to contact Mr Cobbledick and clarified his intentions, assumes that the council would have persisted with a course that was manifestly unreasonable.

[39] For these reasons, neither of the conclusions in para 21 of the Secretary of State's final decision letter can be sustained.

[40] For the sake of completeness, I should mention that Mr Strachan F advanced one further reason as to why it would not be appropriate to award the claimant costs in the present case. The contention is that Mr Cobbledick, in appealing against the enforcement notice, was simply concerned to ensure that upon the quashing of the enforcement notice, he would be entitled to compensation under the stop notice. As the Secretary of State said, his entitlement to compensation under the stop notice is G a separate question. Once that is regarded as a separate question, the question for the Secretary of State was (see para 21 of the decision letter): did the council act unreasonably in issuing the second enforcement notice alleging a material change of use, and had they contacted Mr Cobbledick, might they have acted differently?

[41] There is only one answer to both of those questions: yes. The point H is taken that if the appeal was pursued for the sole reason of claiming compensation in respect of the stop notice, it would not be appropriate to award costs, because they could be claimed as part of the compensation claim in respect of the stop notice. I am not convinced that that is necessarily the case. A claimant for compensation must have mitigated its

loss, and it might be argued that if Mr Cobbledick was entitled to make a claim for costs, he should have done so.

[42] There is a perfectly valid claim for costs to be determined by the Secretary of State. Although the information provided in Mr Cobbledick's written representations for the purposes of the enforcement notice appeal indicated that he had obtained a licence of the premises for that summer season, the extent to which he either had, or had hoped to, obtain a further interest in the property at the time of the appeal against the enforcement notice, was not raised by the Secretary of State as a possible reason for not awarding costs. I simply record that Mr James Maurici, on behalf of the claimant, told me upon instruction that as at the time Mr Cobbledick appealed against the enforcement notice, and that plainly is the relevant time, he was hoping to obtain a further interest in the property. In any event, as at that time he still had a licence to trade in the property for the remainder of the summer season. If the Secretary of State is to take this new point, plainly the claimant must be given a fair opportunity to explain what the position had been when he appealed against the enforcement notice.

[43] The material fact, it seems to me, is that the second enforcement notice did not simply relate to the temporary structure; the council chose to serve the second enforcement notice in respect of the whole of no 14, prohibiting retail use in respect of the whole of the premises. There is no indication that it gave any consideration as to whether an enforcement notice in such broad terms was justified on the merits, given that it did not appear to have an objection in principle, but was simply concerned about detailed matters of design. In any event, the question is whether the claimant, at the time of the appeal against the enforcement notice, was justified in appealing against such a notice.

[44] For all of these reasons, I am satisfied that the decision dated 4 July 2003 must be quashed.

Application allowed.

R (on the application of Cherwell District Council) *v* First Secretary of State and another

[2004] EWCA Civ 1420

COURT OF APPEAL
Chadwick, Dyson LJJ and Munby J
28 October 2004

Planning permission – Proposed development on Crown land – Circular 18/84 – Control of development of Crown land – Part XIII of Town and Country Planning Act 1990 – Use of private contractor on Crown land – Whether development by private contractor development by Crown – Whether notification procedure available – Whether Crown needing planning permission

The government proposed the development of a centre for asylum seekers on land owned by the Ministry of Defence. The Home Office proceeded on the basis of the non-statutory consultation procedure within the memorandum accompanying Circular 18/84 (Crown Land and Crown Development), which contains the notice of proposed development (NPD) procedure. Notice of the proposed development was given to the appellant council in accordance with that procedure. The council objected to the proposals and requested the respondent Secretary of State to hear all objections by way of a public inquiry. At the non-statutory inquiry, the council contended that, because the development would be carried out by a private sector contractor, which would make the reserved matters application and operate the proposed asylum centre, the NPD procedure could not be used; there was evidence before the inspector that the proposals would be achieved by using a design, build and operate (DBO) contract with a private sector contractor. The appointed inspector concluded that the NPD procedure was appropriate, but he recommended that approval should not be given. Contrary to that recommendation, the Secretary of State approved the proposed development. The council sought judicial review of the Secretary of State's decision, on the ground, *inter alia*, that he had adopted the wrong procedure in concluding that the use of the NPD procedure was appropriate in circumstances where the statutory procedure under section 299 of the Town and Country Planning Act 1990 was applicable. That application was dismissed in the court below, Collins J having decided that the development to be carried out by the DBO contractor would fall within section 294(1) of the 1990 Act. On the council's application, permission was given to appeal on the issue of the use of the NPD procedure.

Held The appeal was dismissed.

Development by a DBO contractor, carried out by virtue of a private interest in Crown land, can fall within section 294(1) of the 1990 Act: see [51]. Such development would be immune from enforcement action.

The proposed development would be development by the Crown: see [58]. The NPD procedure was therefore applicable. It was not an abuse of power for the Home Office to use it for the construction of accommodation centres in former defence land: see [67].

Cases referred to in the judgment

Cozens v *Brutus* [1973] AC 854; [1972] 3 WLR 521; [1972] 2 All ER 1297; (1972) Cr AppR 799, HL

Hillingdon London Borough Council v *Secretary of State for the Environment* unreported 30 July 1999, [1999] PLSCS 215

Lord Advocate v *Dumbarton District Council; Lord Advocate* v *Strathclyde Regional Council* [1990] 2 AC 580; [1989] 3 WLR 1346; [1990] 1 All ER 1

Ministry of Agriculture, Fisheries and Food v *Jenkins* [1963] 2 QB 317; [1963] 2 WLR 906; [1963] 2 All ER 147; (1963) 186 EG 179, CA

Moyna v *Secretary of State for Work and Pensions; sub nom Moyna* v *Secretary of State for Social Security* [2003] UKHL 44; [2003] 1 WLR 1929; [2003] 4 All ER 162

Newbury District Council v *Secretary of State for the Environment; Newbury District Council* v *International Synthetic Rubber Co Ltd* (1977) 75 LGR 608; 35 P&CR 170; 242 EG 377; [1977] JPL 373

R (on the application of Cherwell District Council) v *First Secretary of State* [2004] EWHC 724 (Admin); [2004] 2 P&CR 27

R v *Monopolies and Mergers Commission, ex parte South Yorkshire Transport* [1993] 1 WLR 23; [1993] 1 All ER 289

R v *Portus, ex parte Federated Clerks Union of Australia* (1949) 79 CLR 428

R v *Worthing Borough Council, ex parte Burch* (1983) 50 P&CR 53; [1984] JPL 261, QB

Appeal against a decision of Collins J

This was an appeal by Cherwell District Council against an order made by Collins J, on 6 April 2004, on an application to quash the decision of the First Secretary of State to approve proposals by the Home Office for the construction of an accommodation centre for asylum seekers.

David Elvin QC and Reuben Taylor (instructed by Sharpe Pritchard, acting as agent for Cherwell District Council) appeared for the appellant council.

Philip Sales and Sarah Moore (instructed by the Treasury Solicitor) represented the respondent, the First Secretary of State.

Keith Lindblom QC and Rupert Warren (instructed by Berwin Leighton Paisner) appeared for the interested party, the Secretary of State for the Home Department.

The following judgments were delivered.

CHADWICK LJ:

[1] This is an appeal from an order made on 6 April 2004 by Collins J on an application to quash the decision of the First Secretary of State to approve proposals by the Home Office for the construction of an accommodation centre for asylum seekers on land at Piddington, near Bicester. The appellants, Cherwell District Council, are the local planning authority. The appeal raises points of some general importance in relation to the use of the non-statutory procedure introduced by the Department of

the Environment Circular 18/84 (Crown Land and Crown Development) A
in connection with what may loosely be described as public/private partnership development projects.

Proposed development

[2] In February 2002, the government published its proposals for the delivery of "a properly managed, robust and integrated system of B
immigration, nationality and asylum" in a white paper, *Secure Borders, Safe Haven: Integration with Diversity in Modern Britain.* In a letter to the district council dated 14 May 2002, the Minister of State at the Home Office explained that the establishment of accommodation centres for asylum seekers was "a key element of this more tightly managed asylum system". He wrote:

C

> The centres will be an integral part of a seamless end-to-end process, from initial application through to integration or removal. The government will establish new centres, with a total capacity of 3,000, to accommodate a proportion of new asylum seekers from application through initial decision and any appeal. This will be taken forward on a trial basis.
>
> The centres will provide full-board accommodation. Services, including health care, education, interpretation and opportunities for purposeful D
> activities will also be provided for. …

[3] One of the sites selected by the Home Office in furtherance of this policy was redundant land near Bicester, comprising some 13.9ha owned by the Ministry of Defence and formerly used as a defence storage and distribution centre (DSDC). A planning supporting statement, prepared by E
planning consultants on behalf of the Home Office and sent to the district council with the letter of 14 May 2002, describes, in detail, the proposals for the construction of an accommodation centre for 750 asylum seekers on that site. Paragraph 1.3 of that statement explains that:

> The Home Office wishes to secure the planning principle for the development of the site and there will then be a design, build and operate (ie DBO) tender F
> for all the Accommodation Centres. The Home Office will retain ownership of the site but the Centre will be operated by the chosen contractor (ie the successful DBO bidder). The Home Office will work with the chosen contractor on the detailed design for the Centre. …

Statutory control of development

[4] Statutory control of development is now contained in the Town and G
Country Planning Act 1990 (the 1990 Act). Section 55(1) of that Act defines "development" as including the carrying out of building operations in, on, over or under land or the making of any material change of use of any buildings or other land. There is no dispute that, to the extent that the 1990 Act applies to the development of land by or on behalf of the Crown, the Home Office proposals would constitute "development" within the H
meaning of section 55(1).

[5] Section 57(1) of the 1990 Act provides that, subject to the following provisions of that section (which are not material in the present context), planning permission is required for the carrying out of any development of land. Further provisions in Part III of the Act (Control over Development)

provide for the grant of planning permission; the matters to be taken into account in determining applications for planning permission; the right of appeal to the Secretary of State; the power of the Secretary of State to direct that an application be referred to him; and the reference of matters by the Secretary of State to a planning inquiry commission.

[6] Part VII of the 1990 Act contains provisions for the enforcement of planning control. Section 171A provides that the carrying out of development without the required planning consent constitutes a breach of planning control; section 172 empowers the local planning authority to issue an enforcement notice where it appears to them that there has been a breach of planning control; section 179(1) provides that, where there has been a failure to comply with an enforcement notice, the person who is then the owner of the land is in breach of the notice; and section 179(2) provides that an owner of land who is in breach of an enforcement notice is guilty of an offence. Further enforcement powers are contained in sections 183 to 187 (stop notices) and 187B (powers to restrain breaches of planning control by injunction).

[7] Section 106 of the 1990 Act (as substituted by section 12(1) of the Planning and Compensation Act 1991), read with sections 106A and 106B, contains provisions for the control of development through the acceptance of "planning obligations". Section 106 provides that any person interested in land in the area of a local planning authority may, by agreement or otherwise, enter into an obligation restricting the development or use of the land in any specified way, requiring specified operations or activities to be carried out in, on, under or over the land or requiring the land to be used in any specified way: see paras (a)-(c) of section 106(1). An obligation into which the landowner (or other person interested in the land) has entered pursuant to section 106 (a planning obligation) is enforceable by the local planning authority against successors in title (see section 106(4)); and may be enforced by injunction: see section 106(5). Sections 106A and 106B contain provisions for the discharge or modification of planning obligations, including provisions for an appeal to the Secretary of State from the decision of a local planning authority that the planning obligation will continue to have effect without modification.

Control of development on Crown land

[8] The extent to which the statutory control of development is applicable to the development of Crown land is now prescribed by the provisions in Part XIII of the 1990 Act. But those provisions must be read in the light of the underlying principle, recognised by this court in *Ministry of Agriculture, Fisheries and Food* v *Jenkins* [1963] 2 QB 317 (*MAFF*) and affirmed by the House of Lords in *Lord Advocate* v *Dumbarton District Council* [1990] 2 AC 580, that, in the absence of express provision or necessary implication, the Crown is not bound by the Planning Acts.

[9] The position was explained by Lord Denning, Master of the Rolls, in *MAFF*, at p325:

> Looking at the whole of the Town and Country Planning Act, 1947, I am satisfied that the Crown does not need to get planning permission in respect

> of its own interest in Crown lands. The reason it is exempt is, not by virtue of A
> any provision in the Act itself, but by reason of the general principle that the Crown is not bound by an Act unless it is expressly or impliedly included.

That approach was endorsed by Lord Keith of Kinkel (with whose speech the other members of the House agreed) in *Lord Advocate.* He observed, at p602G, that the provisions in section 253 of the Town and
Country Planning (Scotland) Act 1972, read as a whole, made it "quite B
clear that the whole Act proceeds on the assumption that the Crown is not subject to any requirement of planning permission for development carried out by it". He rejected the view, which had prevailed in the Court of Session, that where the Crown is not expressly bound "it is not bound by necessary implication when acting within its rights but is
so bound when acting without any right", holding, at p604C-D, that it C
was preferable:

> to stick to the simple rule that the Crown is not bound by any statutory provision unless there can somehow be gathered from the terms of the relevant Act an intention to that effect. The Crown can be bound only by express words or necessary implication.

D

[10] As appears from the passage from the judgment of Lord Denning MR that I have set out, the planning legislation relevant to the decision of this court in *MAFF* was contained in the Town and Country Planning Act 1947. Section 87 of that Act included, at subsections (2)(*b*), (3)(*a*) and (5) respectively, provisions now found in sections 296(1)(c), 296(2)(a) and
294(1) of the 1990 Act: E

> 296 .– (1) Notwithstanding any interest of the Crown in Crown land, but subject to the following provisions of this section –
>
> ...
>
> (c) any restrictions or powers imposed or conferred by Part III, VII... or VIII... shall apply and be exercisable in relation to Crown land, to the
> extent of any interest in it for the time being held otherwise than by or F
> on behalf of the Crown.
>
> (2) Except with the consent of the appropriate authority –
>
> ...
>
> (a) no order or notice shall be made issued or served under any of the provisions of [section 172]... in relation to land which for the time being is Crown land

G

> ...
>
> 294 .– (1) No enforcement notice shall be issued [under section 172 of the 1990 Act] in respect of development carried out by or on behalf of the Crown after 1st July 1948 on land which was Crown land at the time when the development was carried out.

In that context, "Crown land" includes "land in which there is... an H
interest belonging to a government department": see section 293(1) of the 1990 Act and, formerly, section 87(1) of the 1947 Act.

[11] The effect of the provision now enacted as section 294(1) of the 1990 Act was explained by Robert Goff J in *Newbury District Council* v *Secretary of State for the Environment* (1977) 35 P&CR 170, at p180:

> The subsection provides that no enforcement notice shall be served in respect of such a development. It does not provide, as it easily could have done if such had been the intention, that no enforcement notice should be served *on the Crown* in respect of such development. It follows that, if a private citizen should subsequently acquire any such land from the Crown, he would not have to apply for planning permission in respect of development within the subsection. He too would be protected from the service of an enforcement notice by the terms of the subsection which are quite explicit."

As the editors of the *Encyclopedia of Planning Law and Practice* point out, in note P294.03, at p2-3904:

> Subs(1) [of section 294 of the 1990 Act] confers immunity from enforcement in respect of Crown development, not only for the period that the land remains in Crown ownership (which is in any event unnecessary because of the Crown's immunity) but also thereafter…

Planning consent in anticipation of the disposal of Crown land

[12] The provisions to which I have just referred – section 294(1) of the 1990 Act and its statutory predecessors – protected persons other than the Crown from enforcement proceedings in respect of development carried out by, or on behalf of, the Crown upon land that was Crown land at the time at which the development was carried out, notwithstanding that the land was subsequently transferred away from the Crown's ownership. But those provisions did not, of course, confer protection from enforcement in respect of development carried out on that land by the new owner. And it was the view of the law officers that it was a necessary corollary of the principle that the Crown was not bound by the Planning Acts that the Crown could not itself seek a planning consent in respect of proposed development. The effect was that the Crown could not transfer land to the new owner for development with the benefit of a planning consent that it had obtained. That gave rise to uncertainty in relation to the value to be placed on land that had become surplus to the needs of the Crown for use in the public interest and that the Crown wished to sell on to a private developer. The problem is explained in note P299.03, at p2-3911, in the *Encyclopedia of Planning Law and Practice.*

[13] In an attempt to meet that problem the then Minister of Housing and Local Government prescribed (in Circular 49/63) a procedure under which the disposing department of the Crown and the local planning authority could seek a non-binding opinion from the minister on the question of whether, if an application for development of the land consistent with that which the disposing department considered appropriate had been made and thereafter referred to him on appeal from a refusal by the planning authority, he would or would not have allowed the appeal.

[14] In 1983, that procedure was challenged in *R* v *Worthing Borough Council, ex parte Burch* [1984] JPL 261. It was held that the procedure was unlawful, in that it was intended to constrain the exercise by the local planning authority of their statutory powers when, following the disposal of the land to the private developer, the new owner made a planning application under the Act. The decision in *Worthing* led to the enactment of the Town and Country Planning Act 1984 (the 1984 Act).

[15] Section 1 of the 1984 Act was enacted in these terms, so far as material:

> .– (1) This section has effect for the purpose of enabling Crown land, or an interest in Crown land, to be disposed of with the benefit of –
>
> (*a*) planning permission,...
>
> (2) Notwithstanding the interest of the Crown in the land in question, an application for any such permission... may be made by –
>
> (*a*) the appropriate authority; or
>
> (*b*) any person authorised by that authority in writing;
>
> and, subject to subsections (3) to (5) below, all of the statutory provisions relating to the making and determination of any such application shall accordingly apply as if the land were not Crown land.
>
> (3) any planning permission granted by virtue of this section shall apply only –
>
> (*a*) to development carried out after the land in question has ceased to be Crown land; and
>
> (*b*) so long as that land continues to be Crown land, to development carried out by virtue of a private interest in the land;

Those provisions have been re-enacted as section 299(1)-(3) of the 1990 Act.

[16] The enlarged provisions for the control of development on privately owned land through the acceptance by the landowner of planning obligations introduced by the Planning and Compensation Act 1991, and now contained in sections 106, 106A and 106B of the 1990 Act, were extended to Crown land by section 12(3) of the 1991 Act. That section inserted a new section – section 299A – immediately after section 299 of the 1990 Act. Section 299A(1) enables "the appropriate authority" in relation to any Crown interest in land in the area of a local planning authority to enter into a planning obligation: that is to say, an obligation falling within paras (a)-(d) of section 106(1) of the 1990 Act. In that context, a government department is the appropriate authority in respect of land that it owns: see section 293(2) of the 1990 Act. A planning obligation entered into under section 299A is enforceable by the local planning authority against any person with a private interest in the land deriving from the relevant Crown interest – see section 299A(3) – and may, subject to the consent of the department that is the appropriate authority, be enforced by injunction: see section 299A(4) and (5).

Circular 18/84

[17] The 1984 Act received Royal Assent on 12 April 1984. It was to come into force at the end of the period of four months beginning on that day: see section 7(2). Circular 18/84 was issued by the Department of the Environment on 3 August 1984. Paragraph 1 of the circular is in these terms, so far as material:

> 1. The Town and Country Planning Act 1984 comes into force on 12 August 1984. A commentary on its provisions in relation to England and Wales, and with reference to the regulations made under sections 1 and 3 of the Act which also come into force on the same day, is contained in Part I of the memorandum accompanying this circular. Section 1 of the Act enables

A planning permission to be sought prior to the disposal of Crown land or an interest therein. It thus renders redundant the procedure set out in Ministry of Housing and Local Government Circular 49/63 by which disposing Departments sought an informal opinion from local planning authorities about future uses of surplus Crown land. Circular 49/63 is hereby cancelled. …

B Paragraph 4 described the scope of Part IV of the accompanying memorandum:

> 4. Part IV sets out the arrangements by which Government Departments and other Crown bodies consult local planning authorities about their proposals for development. …

C [18] Part IV of the memorandum includes the following paragraphs:

> **Scope of consultation**
>
> 4. Development by the Crown does not require planning permission. But Government Departments will consult local planning authorities before proceeding with development (including material changes of use) which would otherwise require planning permission. …
>
> D …
>
> **Method of consultation**
>
> 9. When the formal stage of consultation is reached, the developing Department will send to the local planning authority four copies of a statement of their proposal marked "Notice of Proposed Development by (Department)" sufficient to enable the authority to appreciate its nature and extent. …
>
> E …
>
> **Action by the local planning authority**
>
> 12. The local planning authority will treat the Notice in the same way as they would a statutory planning application…
>
> …
>
> **Publicity**
>
> 18. Although development proposals of Departments are not subject
> F to statutory publicity, local planning authorities should give them the same publicity by way of press advertisement, site notices, notification of neighbours etc, as they would if the development proposals were the subject of an application for planning permission. …
>
> **Notification of local planning authorities views**
>
> 19. The local planning authority's views on the proposed development should be sent to the developing Department within eight weeks from
> G the date of receipt of the Notice of Proposed Development… In formulating their views, the local planning authority should have regard to the views of any bodies consulted and to any comments received by the authority in response to the publicity given. They should make it clear whether they find the development acceptable and, if so, on what conditions (if any); or whether they find it unacceptable. In either case,
> H reasons for the conditions or for finding the proposal unacceptable should be clearly stated. …
>
> …
>
> **Method of dealing with objections…**
>
> 23. Where the local planning authority object to the Notice of Proposed Development… or where there is any unresolved disagreement, the developing Department, if they wish to proceed with the proposal, will

> notify the Department of the Environment, Transport and the Regions submitting copies of relevant documents and plans. ...
>
> ...
>
> 25. The method of dealing with proposals to which there have been objections will depend upon the circumstances but in all cases the local planning authority, the developing Department and other interested parties will be given an opportunity to express their views. It is expected that the written representations method will be suitable for most cases. ... In other cases, it may be desirable to take more formal steps and to hold a non-statutory public inquiry. The result of the Secretary of State's determination of the dispute between the developing Department and the local planning authority will subsequently be notified by the Department of the Environment, Transport and the Regions by means of a letter to the local planning authority. ...

References in paras 23 and 25 (and elsewhere in the circular) to the Department of the Environment and to the Secretary of State are now to be read as referring to the Office of the Deputy Prime Minister (ODPM) and to the First Secretary of State: see article 16(5) of the Transfer of Functions (Transport, Local Government and the Regions) Order 2002 (SI 2002/2626).

Use of the non-statutory procedure in the present case

[19] The Home Office, as the developing department, took the view that the present case was one in which it was appropriate to use the non-statutory consultation procedure prescribed in Part IV of the memorandum accompanying Circular 18/84 (the notice of proposed development (NPD) procedure). Notice of Proposed Development was sent to the district council, as local planning authority, on 14 May 2002, with the letter from the minister and the planning supporting statement to which I have already referred. The proposed development was described in the notice in these terms:

> **PROPOSED DEVELOPMENT:**
> Demolition of existing buildings and erection of a Home Office – Immigration and Nationality Directorate – Accommodation Centre – Outline Proposal utilising existing access from the public highway.
>
> **AT:** Land at DSDC, Bicester, Oxfordshire
>
> **RESERVED MATTERS:** siting, design, external appearance and landscaping.

[20] By a letter dated 16 July 2002, the council's head of planning and development services notified the Home Office that, at a meeting on 11 July 2002, the council's south area planning committee had resolved to object to the proposals in the NPD on the five grounds that he had set out. The letter included a request that the ODPM be asked to ensure that any subsequent appeal against this objection be dealt with at a public inquiry "in order to give all interested parties the opportunity to present their concerns in full to an independent inspector".

[21] The matter was referred to the ODPM in accordance with the provisions of para 23 of the NPD procedure. On 9 August 2002, the First Secretary of State notified the council that he had decided to hold

a non-statutory public local inquiry into the matter, so that he could "consider all the relevant aspects of the proposed development". That was a course open to him under para 25 of the NPD procedure.

[22] The inquiry opened on 10 December 2002 and extended over 14 days, concluding on 3 March 2003. The inspector appointed to conduct the inquiry reported to the First Secretary of State on 18 April 2003. He recommended that the proposals should not be approved. The reasons that led him to make that recommendation, on "the overall balance", are summarised in para 14.75 of his report. Although that paragraph has been set out in the judgment now under appeal, it is convenient to include it here:

> In making the overall decision it is my view that greater weight should be attached to the factors against the proposal than for it. I say this for three reasons. First, the need to create a more sustained pattern of development lies at the very heart of the Government's well-established approach to land use planning in this country. It would, in my view, be inappropriate for the Government to make an exception in respect of its own development. Second, the necessity for, and the benefits of, the proposed Accommodation Centre being in this rural area were not, in my view, shown by the Home Office to be convincing or overriding. The proposal seems to me to be driven by the availability of some surplus Government-owned land and by an undemonstrated belief that this rural location would aid the self-containment of the Centre. Third, the risk of road accidents involving pedestrian asylum seekers and others is, in my view, a very real practical worry that was given insufficient attention by the Home Office. My overall conclusion is, therefore, that, on balance, the First Secretary of State should not give approval to the proposed development.

[23] The First Secretary of State did not accept the inspector's recommendation. He approved the proposed development. His decision was communicated to the district council by a letter dated 18 August 2003. The letter contained a detailed explanation of the reasons that led the First Secretary to his decision, and, in para 49, set out the conditions subject to which approval was given. Those conditions were in the same terms as the development conditions set out in schedule 1 to an agreement made pursuant to section 299A of the 1990 Act on 14 April 2003 (shortly after the completion of the inquiry hearings) between the district council, Oxfordshire County Council, the Secretary of State for Defence and the Secretary of State for the Home Department. Notwithstanding the absence of any express power in the First Secretary of State to impose conditions under the NPD procedure, it has not been suggested that the imposition of conditions vitiates his decision.

Challenge to the use of the NPD procedure

[24] In the course of his report, the inspector had addressed a submission, advanced on behalf of the council, that the proposed development could not lawfully be approved by the First Secretary of State under the NPD procedure. He summarised the arguments in support of that submission in paras 7.4 and 7.5 of his report:

Procedural Matters (CDC)

...

7.4 The use and application of the Circular 18/84 procedure by the Home Office has been wrong. It is difficult to understand why the Home Office have not used the procedure set out in Section 299 of the 1990 Act. This contains a specific power which would, notwithstanding the interest of the Crown in the land, allow the Crown to apply for planning permission in anticipation of the grant of the lease to the private contractor selected to implement the proposal. This would have allowed the statutory procedure to be followed and the implementation of planning conditions in the usual way. The difficulty with the non-statutory procedure being followed is that it does not provide the same enforcement mechanisms. Where the section 299 procedure is available, as it is here, there can be no proper reason for the Crown taking advantage of the non-statutory procedure.

7.5 It is difficult to see how the development can be regarded as "Development by a Government Department" within Part IV of Circular 18/84. The "design, build and operate" contractor would not simply be the architect or builder of the Centre for the Home Office but would be independent. The reserved matters Notification to be made would not relate to Crown development but to private sector development. The contractor cannot in fact use the procedure in Circular 18/84 since the contractor would not be the Crown. The Home Office may not itself make the reserved matters Notification since they have no genuine intention to develop the Site themselves. Any such Notification would be an abuse of power in public law. The operator would have to apply for a new full permission to be able to lawfully develop the Site under the contract let by the Home Office. The Circular procedure is not intended to be a means by which the statutory procedure can be avoided, but provides the necessary procedure where the 1990 Act does not apply. Where the 1990 Act does apply, as here, it is neither reasonable nor fair to use Circular 18/84. It is therefore submitted that the First Secretary of State should not grant his approval on the Notification because the wrong procedure has been used and the effect of that error would be to subvert the proper statutory planning control and the public interest. In view of these factors it is seriously in doubt as to whether the First Secretary of State can lawfully approve the development under the Notification procedure.

[25] Among the documents before the inquiry was a draft of the section 299A agreement. The agreement had been executed by the date upon which the inspector made his report. Clause 4.1 of the section 299A agreement provides for the transfer of the development site from the Ministry of Defence to the Home Office "subject to and as soon as reasonably possible after the satisfaction of the First Condition". The "First Condition", in that context, is the approval of the proposed development by the First Secretary of State under the NPD procedure. Clause 5.1.1 provides that the Home Office shall not permit the commencement of the development unless and until: (i) the first condition has been satisfied and the site has been transferred pursuant to clause 4.1; (ii) the DBO contract has been awarded; and (iii) the site has been leased to the DBO contractor for a term that has commenced. Clauses 5.1.3 and 5.1.4 are in these terms:

5.1.3 The Home Office shall include in the DBO Contract and the lease... a specific term that the Contractor will comply with the Planning Obligations

> including the Development Conditions and the Home Office shall not discharge or release (or agree to discharge or release) this requirement or allow it to be treated as discharged otherwise than by performance.
>
> 5.1.4 In addition to the requirement of clause 5.1.3 the Home Office shall use all reasonable endeavours through and under its management of the DBO Contract to ensure the Contractor takes all necessary steps to comply with the Planning Obligations including the Development Conditions throughout the term of the DBO Contract...

The "Planning Obligations" include the "Development Conditions", and include, also, obligations in respect of bus routes, highways, access and car parking. The development conditions are set out in schedule 1 to the agreement. It was contemplated that they were to be the conditions upon which approval would be given by the First Secretary under the NPD procedure: see clause 6.1 of the agreement. As I have said, that is what happened. Effect is given to the conditions set out in the approval letter of 18 August 2003 through the obligations undertaken in the section 299A agreement.

[26] The inspector had before him, also, the evidence of Mr Jonathan Payne, a deputy director at the immigration and nationality directorate (IND) of the Home Office and the project manager for the development of accommodation centres for asylum applicants. Mr Payne's proof of evidence, dated 8 November 2002, included the following description of the basis upon which the DBO contract would be let:

> **Procurement of Accommodation Centres**
>
> 2.1 Accommodation Centres will be procured on a design, build and operate basis. This means that the Home Office will let a contract to a single contractor or consortium via an open competition in accordance with EU procurement regulations. The successful contractor will design, build and operate the centre on the basis of specifications and standards laid down by the Home Office. The contractor may sub-contract some of the works and services to appropriate specialists. The contract will take account of any planning conditions which are imposed.
>
> 2.2. Management of the contract will be the responsibility of IND. The contract manager will be a senior member of IND and will be based permanently on the site, along with a number of support staff. The freehold of the Notification site will be owned by the Home Office, which will assign a lease to the contractor for the duration of the contract.
>
> 2.3 Whilst many if not most of the specifications have been taken into account by the Home Office in developing the proposals at Bicester, plainly there are many points of detail which remain to be finalised as part of the negotiated procurement process. However it is not expected that the elements discussed in my evidence will change, since they emerge from the Government's policy itself.

[27] The inspector referred to the section 299A agreement in para 6.3 of his report in the course of summarising the case advanced at the inquiry on behalf of the Home Office:

> 6.3... Moreover, a detailed section 299A Agreement has been entered into with the Council and this effectively defines the development. It sets out development conditions and operational specifications which

would be binding. The Council would retain the normal level of control of matters at the detailed stage. It cannot be argued that there is insufficient information for a decision to be made or that the section 299A agreement is inadequate to define the development and to regulate and ensure its proper implementation.

[28] In the paragraphs immediately following para 6.3, the inspector set out the response of the Home Office to the district council's submission that the NPD procedure should not have been used:

> **Procedural Matters** (HO)
>
> 6.4 The Council argue that the Notification cannot lawfully be approved by the First Secretary of State. This is on the basis that while the Home Office have applied for clearance from the First Secretary of State it would be a private sector contractor who would make the reserved matters application, and build and operate the Centre, rather than the Crown itself, and that the Circular 18/84 Notification procedure cannot be used for a private sector development, even one on Crown land. The Council argue that in any case where section 299 of the Town and Country Planning Act 1990 (the 1990 Act) offers the Crown the opportunity of making a planning application, to which planning conditions may be attached, it is an abuse of power to seek to use the Notification procedure.
>
> 6.5 These arguments are, however, misconceived and should be rejected. The Home Office would retain a close degree of control over the "design, build and operate contract" through the contract management and lease. Such control is enforceable under the Section 299A Agreement by the Council as a matter of contract. The development would be run throughout under the auspices of the Home Office, who would gain a direct benefit from it and they would fund it in partnership with the contractor. The Notification process is a means of consulting upon the clearance of consent for an act of development. That does not refer to the operation of the Centre but to the initial act of development, which is to be carried out by the contractor for, and under the control of, the Home Office.
>
> 6.6 The facts here can only support a finding that this is a proposed development by a Government department. There can be no proper objection if the contractor submits the reserved matters application. It would still be Crown development.
>
> 6.7 The Council's argument that, where section 299 of the 1990 Act gives the Crown the option to make a planning application, it is an abuse of power not to do so, is bad. The terms of the Section 299A Agreement are such that there is little practical difference between the enforcement of the Notification proposal and the conditions that would be attached to a Section 299 consent. The ability of the Council to take enforcement action has its counterpart in the provisions of the Agreement. The Agreement itself may be varied by the parties with at least as much flexibility as the variation of conditions. The Council's submission that the Home Office is involving itself in, or inviting an abuse of power, should therefore be rejected...

[29] The inspector's conclusion on the procedural challenge raised by the district council is found in para 14.4 of his report:

Procedural Matters

> ...
>
> 14.4 Whether or not the Home Office use and application of the Circular 18/84, Crown Land and Crown Development, procedure (which does not involve the grant of planning permission) in this case is correct is a matter of law for the First Secretary of State. The Council maintain that it is not and argue that any approval by the First Secretary of State would be an abuse of power. In my view, however, the Home Office are entitled to use either the Circular 18/84 procedure or the procedure in Section 299 of the Town and Country Planning Act 1990 (which does involve the grant of planning permission) as a matter of choice; there is nothing in the legislation to say otherwise. Nor does there appear to me to be any legal problem, as the Council believe, about who makes the detailed reserved matters proposal in due course. I consider that the development would clearly be Crown Development and that either the Home Office or the private contractor appointed to design, build and operate it could make the reserved matters proposal under the Circular procedure. Permissions and approvals generally run with the land and there is no reason to treat this case differently.

[30] On that issue, the First Secretary of State agreed with the inspector. He addressed the council's submission that the Home Office had been wrong to use the NPD procedure in this case in para 5 of his letter dated 18 August 2003. He wrote:

> **Procedural Matters**
>
> **Application Process**
>
> 5. At the inquiry, Cherwell District Council said that the use and application of Circular 18/84 – *Crown Land and Crown Development* – was wrong and that the Home Office should have used the procedure set out in Section 299 of the Town and Country Planning Act 1990 ("the 1990 Act"). For the reasons given by the Inspector in paragraph 14.4 of his report, the Secretary of State agrees that it was open to the Home Office to use either the provisions of Circular 18/84 or section 299 of the 1990 Act. He agrees with the Inspector that the proposal is clearly development by the Crown and that either the Home Office or the private contractor appointed to design build and operate could make a reserved matters proposal under the Circular 18/84 process.

These proceedings

[31] These proceedings were commenced by the issue of a claim form on 31 October 2003. The council sought judicial review of the decision of the First Secretary of State, contained in the letter of 18 August 2003, to which I have referred, to approve the development proposed in the notice dated 14 May 2002. The seven grounds of challenge were set out in para 31 of a lengthy and detailed statement attached to the claim form. In the present context, it is necessary only to note the first of those grounds: that the First Secretary had erred in law in his consideration of "the use of the Circular 18/84 procedure in circumstances where the statutory procedure under s299 TCPA was applicable". The relief sought included: (i) a declaration that "the use of the Circular 18/84 procedure was unlawful and that the Home Office should if it intends to proceed be required to submit a planning application for the proposed Accommodation Centre under section 299 of the [1990 Act] and thereafter follow the statutory

procedure for the determination of such an application"; and (ii) an A
order quashing the decision contained in the letter of 18 August 2003 and remitting the matter to the First Secretary of State for him to reconsider according to law.

[32] The respondent to the application for judicial review was the First Secretary of State, whose decision was the subject of the challenge, but the Secretary of State for the Home Department was joined as an B
interested party. Mr Payne, the project manager for the development of IND accommodation centres, gave evidence in a witness statement dated 10 February 2004. He disclosed in that witness statement that a "preferred bidder" for the proposed DBO contract had been selected in January 2004, and that, although the final form of the DBO contract had not been agreed, its principal features had been established. He exhibited: (i) the form of C
contract under negotiation (redacted to exclude commercially sensitive clauses), which, as he said, "would assist the Court in an appreciation of the degree of control exercised by [the Home Office] over the design, build, operate and maintenance provisions of the contract"; and (ii) a form of the lease that it was proposed to grant to the appointed contractor. He explained, in para 7.6 of his statement, that the Home Office "will only D
agree the final contractual terms on the basis that the accommodation centre is developed on its behalf, and it maintains close control and a "step in" power over its operation". In para 7.30, after referring to clause 7 of the proposed lease, he said: "The contractor's leasehold interest in the site is subject to its due performance of its contractual obligation to design, build, operate and maintain an accommodation centre for and on behalf of [the Home Office]". Those statements have not been challenged. They E
led Collins J to conclude – in a passage to which I shall refer later in this judgment – that: "It is clear that the Accommodation Centre will be constructed and operated for the purposes of the Home Office... There is no other purpose behind this development." On the material that was before the judge, that conclusion was inevitable.

[33] Permission to apply for judicial review was granted by F
Richards J on 20 January 2004. The substantive application came before Collins J for hearing in March 2004. On 6 April 2004, he ordered that the claim be dismissed. He reached that conclusion for the reasons set out in the judgment that he handed down on that day: [2004] EWHC 724 (Admin). He refused permission to appeal.

[34] Permission to appeal was granted by this court (Carnwath LJ) on G
20 May 2004, but was limited to ground 1 in the appellant's notice. It is convenient to set out that ground:

> (1) The learned judge erred in law in rejecting the Appellant's first ground of challenge namely that the Secretary of State for the Home Department adopted the wrong procedure in respect of their proposed development. In H
> particular the learned judge erred in rejecting the submissions that (i) the Secretary of State was obliged in the circumstances of the case to use the specific statutory procedure under Part XIII of the Town and Country Planning Act 1990 and (ii) the proposed development was not development by the Crown to which Crown immunity attached since given both the Home Office's proposals and the terms of the legal agreement entered into the

> development could only be carried out by a private developer in possession of a private interest in the land to be developed.

Judge's reasons

[35] The judge addressed each of the seven grounds of challenge raised in the claim before him. But, in this court, it is necessary to consider only that part of his judgment that relates to the matter in respect of which permission to appeal has been given – that is to say, the use of the NPD procedure in this case.

[36] At [17] of his judgment, the judge accepted that, in the circumstances, where no development would be carried out until after the lease had been granted to the DBO contractor – so that the contractor would be carrying out the development by virtue of its private interest (as leaseholder) in the land in question – it would have been open to the Home Office to apply for planning permission under the provisions of section 299 of the 1990 Act. That, I think, is not in dispute and, as it seems to me, is plainly correct. It could have been said that an application for planning permission in respect of the proposed development of the site was made for the purpose of enabling a leasehold interest in the site to be granted to the DBO contractor with the benefit of planning permission. But, as the judge appreciated, the fact that it would have been open to the Home Office to apply for planning permission under section 299 did not of itself lead to the conclusion that the department was bound to do so. The judge went on to say:

> But it is apparent that s299 was enacted to enable the Crown to have an advantage that it otherwise would not have, namely the ability to dispose of land with the benefit of planning permission. Section 299(2) uses the word "may". I see no reason to suppose that Parliament intended that, whenever the Crown wished to dispose of land or an interest in land with a view to development being carried out by a private person, s299 had to be used.

That, if I may say so, is also plainly correct.

[37] If planning permission had been sought and obtained by the Home Office under section 299 of the 1990 Act, that permission would have applied to the development carried out by the DBO contractor by virtue of its private (leasehold) interest – section 299(3)(b) of the Act – whether or not the development was to be carried out by, or on behalf of, the Crown. But the important question, as the judge recognised in [20] of his judgment, was whether the development that was to be carried out by the DBO contractor by virtue of its private interest as leaseholder would be "development carried out by or on behalf of the Crown". Notwithstanding the grant of the leasehold interest, the land would remain Crown land: section 293(1). If the development was carried out on Crown land on behalf of the Home Office, section 294(1) would prevent the issue of an enforcement notice. In practice, therefore, and provided that that condition were satisfied, the DBO contractor could carry out the development without planning permission. It would be protected from enforcement proceedings.

[38] The judge addressed the question as to whether the proposed development was to be carried out on behalf of the Crown in [20] to [23] of his judgment. After referring to the decision of Forbes J in *Hillingdon London Borough Council* v *Secretary of State for the Environment* CO/1604/98, CO/1745/98 and CO/1770/98, unreported 30 July 1999, the judge said: A

> 22. It is clear that the Accommodation Centre will be constructed and operated for the purposes of the Home Office. The precise manner in which the centre is brought into being and operated and whether or not a lease is granted and whether before or after any development cannot detract from that. There is no other purpose behind this development. Mr Elvin [Mr Elvin QC, counsel for the District Council] in reply expressed horror at what he submitted was a novel suggestion that if the Crown was the freeholder it could require any development to be carried out by a private individual and such development would be immune from control. But that is not what is being decided. I see nothing in Forbes J's decision that suggests he was not intending to apply a general test. In these days of public private partnerships and the involvement of the private sector in constructing and operating developments which are manifestly for Crown purposes, such as this Centre, the approach Forbes J approves is in my view manifestly correct. I have no doubt that this development will be carried out on behalf of the Crown and so will be within s294(1)... Although the facts of a given case may well dictate whether any development can be said to be by or on behalf of the Crown, in my view the circumstances of this case are such that no other conclusion could reasonably be reached. B C D
>
> 23. It follows that the development by the DBO contractor does not require planning permission since it is development on behalf of the Crown and is protected by Crown immunity... E

[39] In reaching that conclusion, the judge, as he said, followed the approach adopted by Forbes J in *Hillingdon*. Forbes J had considered what meaning should be given to the expression "development carried out by or on behalf of the Crown" in the context of section 294(1) of the 1990 Act in a lengthy passage (extending, in the version of the transcript that is before this court, over pp33-41 of a 72 page judgment). In substance, he accepted the submission made on behalf of the Secretary of State for the Environment (who, as it happens, was represented by Mr David Elvin QC in that case) that the expression "by or on behalf of the Crown" was not to be treated as having the same meaning as "by the Crown, its servants or agents". The proper test, as Mr Elvin had emphasised, was "not whether the person doing the work is the Crown or a servant or an agent of the Crown", but whether the development is "by or on behalf of the Crown" in the "public law sense". Forbes J expressed his conclusion in the two paragraphs that Collins J set out in [21] of his judgment in the present case: F G

> I am satisfied that the expression "by or on behalf of the Crown" is not to be interpreted, whether for the purposes of the common law principle of Crown immunity or for the purposes of Section 294 of the TCPA 1990, by a strict application of the private law definitions of master and servant and principal and agent. I agree with Mr Elvin that a wider and less restricted interpretation of those words is appropriate, at the very least when H

> considering whether, as in the present case, the method whereby the Crown seeks to achieve its purpose in the development of land is "development by or on behalf of the Crown".
>
> Where (as in the present case) the context and subject matter, which is provided by the contractual provisions and other relevant circumstances relating to the development in question, demonstrate a significant degree of control by the Crown over the work being or about to be carried out by the independent contractor and show that the purposes of the Crown, from which it will derive significant benefit, are to be achieved by the particular method which has been adopted for developing Crown land, the essentially factual conclusion, that the development in question is "development by or on behalf of the Crown" may well be justified as in the present case.

[40] His conclusion that the development to be carried out by the DBO contractor in the present case would fall within section 294(1) of the 1990 Act enabled Collins J to hold that, in the events that had happened, the development would not be subject to statutory planning control. But his conclusion on that point did not of itself answer the question as to whether it had been lawful for the Home Office to choose the non-statutory procedure rather than to apply for planning permission under section 299 of the Act. The judge referred to that question in [15] of his judgment: "Mr Elvin QC... submits... that the non-statutory procedure should not have been used"; and, in [16]: "The inspector's view was that... the Home Office was entitled to use either the statutory or the non-statutory procedure. The [First Secretary] agreed with the inspector's opinion."

[41] It is clear that the judge decided that question against the district council. His reasons for the conclusion that he reached on that point are not, perhaps, set out as clearly as those in relation to the other points that he decided, but they are found, I think, in the following passages of his judgment:

> 18. Mr Elvin contends that his submission [that the NPD procedure should not have been used] is supported by the lesser powers of enforcement which apply if the Circular is used. The whole panoply of enforcement notices, stop notices and injunction can be used against a private individual but not against the Crown. However, enforcement under Part XIII is subject to special rules... [Reference is made to section 294 of the Act]... Section 299A enables the Crown to enter into planning obligations which echo those referred to in s106(1)(a) to (d). There is such an agreement in this case. Breaches of any of those obligations can be enforced against (*sic*) by injunction, but by s299A(5), the appropriate authority (in this case the Home Office) must consent to any such action. In the agreement it had to offer all reasonable co-operation in this respect and any failure would be enforceable through the court.
>
> 19. Other than in relation to enforcement, it is accepted that the non-statutory procedure has provided the claimants and all other objectors with all the safeguards and rights which they would have been afforded had this been an application for permission by a private developer. If the development is properly to be regarded as having been carried out on behalf of the Crown, s294(1) would apply so that enforcement is no different. ...
>
> ...

> 23. … The existence of the s299A agreement and the 18/84 procedure has given the necessary protection and [the First Secretary] was right to reject the [Council's] arguments. I too reject them.

[42] When those passages are analysed, it can be seen that the judge's conclusion that it had been lawful for the Home Office to choose the non-statutory NPD procedure rather than to apply for planning permission under section 299 of the Act rests on the following propositions, namely that: (i) save as to enforcement, the NPD procedure gives to objectors the same degree of protection (in respect of publicity, consultation and right to object) as the statutory procedure following an application under section 299A; (ii) control through the acceptance of planning obligations under the section 299A agreement in this case is as effective as control through the acceptance of planning obligations under section 106 of the Act; and (iii) if, on a true appreciation of the facts, the proposed development is development on Crown land carried out by or on behalf of the Crown, section 294(1) prevents the issue of enforcement notices against the developer (and the Crown has immunity) *whether or not planning permission has been sought and obtained under section 299 of the Act*. That third proposition is, I think, what the judge had in mind when he said, in [19] of his judgment, that: "If the development is properly to be regarded as having been carried out on behalf of the Crown, s294(1) would apply so that enforcement is no different."

Appeal

[43] Although the points upon which the appellants seek to rely, as they appear in ground 1 of the appellants' notice, are put in the following order – (i) the Home Office was obliged in the circumstances of the case to use the specific statutory procedure under Part XIII of the 1990 Act; that is to say, to apply for planning permission under section 299, and (ii) the proposed development was not development by or on behalf of the Crown to which Crown immunity (or immunity under section 294(1)) attached – they are properly to be considered in the reverse order. That is because, absent immunity or the protection from enforcement afforded by section 294(1) of the Act, the development could not lawfully be carried out without planning permission. In such a case, unless the Home Office had sought and obtained permission under the provisions in Part XIII of the Act (more specifically, under section 299), the developer would have had to apply under Part III: section 62. The choice would not be between the use of the NPD procedure or an application for planning permission under Part XIII: the choice would be between an application by the Home Office under Part XIII or an application by the DBO contractor under Part III.

Development by or on behalf of the Crown

[44] I turn first, therefore, to the question as to whether the proposed development was a development by or on behalf of the Crown.

[45] It is, I think, important to have in mind the particular context in which that question arises for the purposes of this appeal. On analysis, it can be seen that there are three distinct contexts in which the question

has been, or might need to be, considered. First, it was necessary for the Home Office, when deciding whether to invoke the NPD procedure or to apply for planning permission under section 299 of the 1990 Act, to consider whether (in the absence of statutory planning permission) the development could be carried out under the protection of section 294(1). Unless the Home Office were satisfied that the development could be carried out by the DBO contractor without statutory planning permission, the department's decision to invoke the NPD procedure could have been said to frustrate a legitimate expectation that statutory planning permission would be sought. Second, it was necessary for the First Secretary of State, when giving approval under the NPD procedure, to satisfy himself that the NPD procedure had been properly invoked. In that context, the First Secretary needed to satisfy himself that the proposals described in the notice of 14 May 2002 and its accompanying documents were for development within para 4 of Circular 18/84 – to which (but to which alone) the procedure set out in Part IV of the accompanying memorandum applied. Unless satisfied that the proposed development was one for which statutory planning permission was not required, the First Secretary could not properly entertain a request by the Home Office, under para 23 of the NPD procedure, to determine the dispute with the local planning authority, and he could not properly proceed to a decision under para 25. Third, the question would (or might) need to be considered by the First Secretary under section 174 of the 1990 Act, and (on appeal from his decision under section 289 of the Act) by the High Court, in the event that, after the commencement of the development, and in the absence of statutory planning permission, the local planning authority were to serve an enforcement notice on the DBO contractor. That was the context in which the question arose before Forbes J in *Hillingdon:* see his description of the issues in appeals A-C under reference CO/1604/98.

[46] On this appeal, the question arises only in the context of the First Secretary's decision, in his letter of 18 August 2003, that: (i) the NPD procedure had been properly invoked; (ii) he could properly entertain a request by the Home Office, under para 23 of the NPD procedure, to determine the dispute with the local planning authority to which the objections to the notice of 14 May 2002 had given rise; and (iii) he could properly proceed to a decision under para 25. There is no formal challenge, in these proceedings, to the decision by the Home Office to issue the notice of 14 May 2002, although, of course, it is inherent in the challenge to the decision of the First Secretary that is the subject of these proceedings that the district council contend that the Home Office decision of May 2002 was wrong. And there is no challenge, in these proceedings, to the issue of an enforcement notice; matters have not reached that stage.

[47] Put shortly, the challenge is to the First Secretary's decision – recorded in para 5 of the letter of 18 August 2003 – that "the proposal is clearly development by the Crown". In reaching that decision, the First Secretary expressed his agreement with the inspector's conclusion (in para 14.4 of his report) that "the development would clearly be Crown development". It is reasonable to infer that he preferred the arguments

advanced by the Home Office – set out in paras 6.4-6.7 of the report to those of the district council: see paras 7.4 and 7.5.

[48] We were urged by counsel for the First Secretary of State to take the view that the question as to whether the proposed development (as presented to the First Secretary under the NPD procedure) was properly to be regarded as a development by or on behalf of the Crown was "a matter of fact and degree which depends on the circumstances of each case", with the consequence (it was said) that the First Secretary's decision on that question was not susceptible to judicial review unless it fell outside the bounds of reasonable judgment. In support of that view, we were referred to observations of Lord Hoffmann in [20], and [25] to [28] of *Moyna* v *Secretary of State for Work and Pensions* [2003] UKHL 44; [2003] 1 WLR 1929, at pp1934-1936.

[49] The point is illuminated by Professor Paul Craig QC in his work *Administrative Law* (5th ed), at pp488-489. In discussing the distinction between law and fact in the context of judicial review, he writes (*ibid*, at p489):

> It [a rigid analytical approach] may lead to the situation in which all inferences drawn from certain primary facts, whether something existed, was present, etc, will be deemed to be questions of law. This would mean that the construction of most statutory terms would be characterised as a question of law. If, in addition, the judiciary then substitute their opinion as to the precise meaning that each of these terms should bear for that of the initial decision maker, the result will be an extensive form of review.
>
> That is not the only possible approach for the courts to adopt. They could interpret the word "law" in a more *pragmatic* or *policy orientated* sense. The application of certain statutory terms might be deemed questions of fact, or the courts could call the application of a statutory term a question of law, but accept that it does not have to have only the one meaning which the court itself would accord to it. Provided that the authority adopts a meaning which is reasonable or has a rational basis, the courts could accept that interpretation, even if it did not accord with the precise meaning which they would have ascribed. …

And, after considering the observations of Lord Mustill in *R* v *Monopolies and Mergers Commission, ex parte South Yorkshire Transport Ltd* [1993] 1 WLR 23, at p32, where the question was whether the jurisdictional condition that the supply of more than 25% of the services of any description "in a substantial part of the United Kingdom" would be carried on by one person had been satisfied, Professor Craig concludes (*ibid*, at p500):

> The court will not necessarily substitute its judgment for that of the agency in circumstances like those in the *South Yorkshire Transport* case. In such a case the reviewing court will define the actual meaning that the statutory term is to have, but where that particular meaning is itself inherently imprecise the court will only intervene if the application of the term is so aberrant as to be irrational.

[50] It is said on behalf of the district council that where (as in this case) development on Crown land is to be undertaken by a private contractor, which is occupying the development site by virtue of its own

A property interest under a lease, the development cannot be development by or on behalf of the Crown. That, as it seems to me, is properly to be characterised as a question of law. If, as a matter of law, development by a DBO contractor carried out by virtue of a private interest in the land cannot be development carried out "by or on behalf of the Crown" for the purposes of section 294(1) of the 1990 Act, the court must say so. On that B question, it must substitute its own view for that of the First Secretary. But if there is no rule of law that development by a DBO contractor carried out by virtue of a private interest in the land cannot be within section 294(1), the question as to whether the proposed development (as presented to the First Secretary under the NPD procedure) was properly to be regarded as a development by or on behalf of the Crown was, indeed, a matter of fact C and degree, and the First Secretary's decision on that question should not be disturbed unless it falls outside the bounds of reasonable judgment.

[51] I am not persuaded that, as a matter of law, development by a DBO contractor carried out by virtue of a private interest in land that is Crown land cannot be within section 294(1) of the 1990 Act.

[52] It is important to keep in mind that the provisions now in D section 294(1) of the 1990 Act – and first enacted as section 87(5) of the 1947 Act – were not enacted for the protection of the Crown: as Robert Goff J explained in *Newbury*, at p180. The Crown does not need the protection of those provisions. The Crown is not subject to planning control under the Planning Acts. Service of an enforcement notice on the Crown would be of no effect. The purpose of the provisions now in section 294(1) of the E 1990 Act is to protect persons other than the Crown from enforcement proceedings. The persons to be protected are persons who would otherwise be affected by the issue and service of an enforcement notice.

[53] It is pertinent, therefore, to read section 294 in conjunction with the provisions relating to the issue of enforcement notices, and the persons affected by such notices. I have referred earlier in this judgment F to sections 171A, 172 and 179 of the 1990 Act. The power to issue an enforcement notice arises where it appears to the local planning authority that: (a) there has been a breach of planning control; and (b) it is expedient to issue the notice, having regard to the provisions of the development plan and to any other material considerations: see section 172(1). A copy of the notice is to be served: (a) on the owner and occupier of the land to G which it relates; and (b) on any other person having an interest in the land, being an interest that, in the opinion of the authority, is materially affected by the notice: see section 172(2). Failure to comply with an enforcement notice is an offence on the part of the owner of the land – section 179(1) and (2) – but subject to the defence under section 179(3), and to carry on any activity that is required by the notice to cease is an offence on the part H of a person (other than the owner) who has control of, or an interest in, the land: see section 179(4) and (5). In that context "owner" means a person entitled to receive the rack rent of the land: see section 336(1) of the Act.

[54] Where the land is Crown land, the person affected by the issue and service of an enforcement notice under section 172 of the Act will be the person (if any) who is entitled to occupy the land by virtue of a private

interest. "Private interest" is defined in section 293(1). It was in order A to enable planning control to be enforced in respect of development on Crown land carried out otherwise than by, or on behalf of, the Crown in a case where no person is entitled to occupy the land by virtue of a private interest that the special enforcement notice procedure now contained in section 294(3) to (7) of the 1984 Act was introduced by section 3 of the 1984 Act. B

[55] It is clear, therefore, that, in a case where the land remains Crown land, the purpose – indeed, the only purpose – of the provisions now found in section 294(1) of the 1990 Act is to give protection from enforcement proceedings to the person (if any) who is entitled to occupy the land by virtue of a private interest. The protection is from enforcement proceedings "in respect of development carried out by or C on behalf of the Crown". To qualify for that protection the land must have been Crown land at the time at which the development was carried out. The existence of the private interest in the land does not prevent land from being Crown land. The fact (if it be the case) that the private interest in the land existed at the time at which the development was carried out does not prevent the land from being Crown land at the D relevant time. A person who is in occupation of the land by virtue of a private interest is not disqualified from the protection of section 294(1) of the Act by reason only of the fact that the private interest existed at the time that the development was carried out. The only relevant question in each case is whether the development was carried out by, or on behalf of, the Crown. E

[56] I am unable to identify any reason in principle why the answer to that question should depend, as a matter of law, upon whether the person carrying out the development has a private interest in the land at the relevant time or upon whether he is able to carry out the development by virtue of his private interest. In my view, Mr Elvin was correct to submit to Forbes J in *Hillingdon* that the expression "by or on behalf of the Crown", F in the context of section 294(1) of the 1990 Act, was not limited to the strict private law concept of master and servant or principal and agent, but had the wider meaning adopted by the High Court of Australia in *R* v *Portus, ex parte Federated Clerks' Union of Australia* (1949) 79 CLR 428, and that that "wider and less restricted interpretation of those words is appropriate, at the very least when considering whether... the method G whereby the Crown seeks to achieve its purpose in the development of land is 'development by or on behalf of the Crown'". In rejecting the local planning authority's submission that "as a matter of law, Crown immunity *only* applies to Crown development of land when the development in question is physically carried out by the Crown itself or by its servants or by its agents, in the strict private law sense of each of those latter H concepts", Forbes J said:

> I accept Mr Elvin's submission that such a restricted interpretation of the expression "by or on behalf of the Crown" would have profound implications for the concept of Crown development of land, is not justified by the natural and ordinary meaning of the words themselves and is not to be preferred to

> the type of approach which was adopted to the meaning and application of such words by the High Court of Australia in the *Portus* case...

Forbes J was correct to take that view. Collins J was right to follow that approach. As he said, in the passage in [22] of his judgment in the present case that I have already set out:

> In these days of public private partnerships and the involvement of the private sector in constructing and operating developments which are manifestly for Crown purposes, such as this Centre, the approach Forbes J approves is in my view manifestly correct.

[57] In any particular case, the existence of the DBO contractor's private interest, and the fact that the development is carried out by virtue of that interest, may well lead to the conclusion that the development is carried out in order to serve the private interest, rather than on behalf of the Crown. But that will be a question of fact and degree, turning on the particular circumstances in each case. Whether that question is properly to be characterised as involving a question of law, as Lord Reid suggested in *Cozens* v *Brutus* [1973] AC 854, at p861, in the passage cited by Lord Hoffmann in *Moyna,* at p1935 it should be seen as one of those questions as to which (to adopt the words of Lord Hoffmann in *Moyna* in [27]) "lawyers have decided that it would be inexpedient for an appellate tribunal to have to form an independent judgment". The court should not substitute its own view unless the view of the decision maker was outside the bounds of reasonable judgment.

[58] In the present case, the First Secretary of State took the view that "the proposal is clearly development by the Crown". That was the view taken by the inspector in para 14.4 of his report. Collins J thought it clear that the accommodation centre would be constructed and operated for the purposes of the Home Office. As he put it: "There is no other purpose behind this development." He took the view that, in the circumstances, of this case, the only conclusion that the First Secretary could have reached was that the development was to be carried out by, or on behalf of, the Crown. I agree.

Notwithstanding the availability of the NPD procedure, should planning permission have been sought under Part XIII of the 1990 Act in the present case?

[59] It is said on behalf of the district council that even if the proposed development was properly to be regarded as development by, or on behalf of, the Crown so as to fall within para 4 of Circular 18/84, it was, nevertheless, an abuse of power for the Home Office to use the NPD procedure when it was open to it to apply for statutory planning permission under section 299 of the 1990 Act. As it is put in the appellants' skeleton argument: "In other words, given the arrangements the HO chose to adopt it was obliged to use the s299 procedure." The basis of that submission is that an application for planning permission under section 299 "would have applied an appropriate statutory procedure and allowed the imposition of conditions in the usual way, which would have

applied to any subsequent occupier of the Notification Site". Further, it is said that the grant of planning permission "would also have allowed the Council to take enforcement action under [the 1990 Act] against any breaches of condition by the DBO contractor or other later occupier of the [accommodation centre]".

[60] The first of those points is difficult to follow and it was not pursued in argument with any enthusiasm. It is, of course, correct that, where an application is made under section 299(1) of the Act, "all the statutory provisions relating to the making and determination of any such application" should apply as if the land were not Crown land: see section 299(2). But it is not suggested that, in the present case, the NPD procedure did not afford safeguards as to notification, publicity, consultation and the right to object in relation to the decision that the First Secretary made in his letter of 18 August 2003 that were the equal of those that would have been afforded under the statutory procedure. As the judge said, in [19] of his judgment: "Other than in relation to enforcement, it is accepted that the non-statutory procedure has provided the claimants and all other objectors with all the safeguards and rights which they would have been afforded had this been an application for permission by a private developer."

[61] Nor, as it seems to me, is it possible to sustain an objection based upon the absence of statutory planning conditions. In the present case, as I have sought to explain, matters that would, or might, otherwise have been the subject of statutory planning conditions – the "Development Conditions" set out in schedule 1 to the section 299A agreement and in the First Secretary's decision letter – were accepted as planning obligations. Those planning obligations will be enforceable by the local planning authority against persons with a private interest deriving from the Crown (including, in this context, the DBO contractor) – see section 299A(3) – to the same extent as if accepted under section 106 of the Act: see section 299A(4). But, whether accepted under section 106 or under section 299A, planning obligations would not be enforceable by injunction or by entry onto the land without the consent of the Home Office: see sections 296(2)(aa) and 299(5). As the judge put it in [23] of his judgment: "The existence of the s299A agreement… has given the necessary protection."

[62] The real complaint made by the district council is that they cannot take enforcement action under the 1990 Act. The point is made in paras 59 and 60 of the skeleton argument:

> 59. The difficulty with the non-statutory procedure is that, while it has certain parallels with the statutory procedures, it does not provide the same enforcement mechanisms and does not protect the public interest as would effective statutory planning control.
>
> 60. In particular, the non-statutory process does not permit the imposition of conditions on an approval which would have the same effect as conditions on a statutory grant of planning permission and which would permit enforcement action pursuant to Part VII of the TCPA 1990. This means that even if conditions, of a type, are imposed on the non-statutory approval they are not conditions in the sense of conditions imposed on a planning permission and breach of them (even by a private individual taking an

> interest in the land from the Crown) would not be subject to the statutory enforcement process under eg ss172, 183 or 187A or 187B.

[63] The fallacy underlying that complaint, as it seems to me, is that it assumes that the statutory enforcement process would be available if planning permission had been sought and obtained under section 299 of the Act. But that assumption overlooks the fact that the reason why a person having a private interest in the land (for example, the DBO contractor) is protected from enforcement proceedings is not because planning permission has not been obtained under Part XIII of the Act (or under Part III, as the case might be), but because the development falls within section 294(1) of the Act. This was the point made by the judge in the sentence in [19] of his judgment to which I have already referred: "If the development is properly to be regarded as having been carried out on behalf of the Crown, s294(1) would apply so that enforcement is no different." Other than to assert that the development is not properly to be regarded as having been carried out on behalf of the Crown, the district council offer no answer to that point.

[64] As I have said, if planning permission had been sought on an application made by the Home Office under section 299(1) of the Act all the statutory provisions "relating to the making and determination of any such application" would have applied as if the land were not Crown land: see section 299(2). Permission, if granted under section 299, would apply to development carried out by the DBO contractor after the grant of the lease: see section 299(3)(b). But section 299 of the Act says nothing about the enforcement of planning control. Enforcement is governed by other provisions in the 1990 Act and (in relation to the Crown) by immunity under the general law.

[65] It was held by this court, in *MAFF,* in a passage at p325 to which reference has already been made, that the Crown does not need to get planning permission in respect of its own interest in Crown lands. The reason is that the Planning Acts – in that case the Town and Country Planning Act 1947 – do not bind the Crown. But the Planning Acts do bind the Crown's tenants in respect of their own interests. That was made clear by section 87(2)(*b*) of the 1947 Act, now section 296(1)(c) of the 1990 Act. As Lord Denning observed, in a passage (*ibid*) upon which the council place much emphasis:

> [Section 87(2)(b)]... proceeds on the assumption that the Crown is already exempt. It says that, "Notwithstanding any interest of the Crown in land being Crown land... any [planning] restrictions... shall apply and be exercisable in relation to the land, to the extent of any interest therein... held otherwise than by or on behalf of the Crown..." That provision assumes that the Crown is already exempt in respect of its own interest in Crown land. All it does is to make sure that other persons (eg, its tenants) have to get planning permission in respect of their interests;...

But section 87(2)(*b*) of the 1947 Act was expressed to be subject to section 87(5) of that Act: now re-enacted as section 294(1) of the 1990 Act. Nothing said by this court in *MAFF* could have been intended to restrict the effect that the provisions formerly in section 87(5) of the 1947 Act, and A now in section 294(1) of the 1990 Act, would otherwise have. The court did not refer to those provisions. In deciding the issue that was before it in *MAFF*, it did not need to do so. They had no relevance in that context.

[66] The Planning Acts apply to private interests in Crown land. That is made clear by section 299(3)(b) and (6) of the 1990 Act. But they apply subject to the provisions in section 294(1) of that Act, and also to B the provisions in section 296(2)(a), which prevent enforcement under sections 172, 183, 187A and 187B in relation to land that, for the time being, is Crown land without the consent of the relevant government department. Those provisions that have been in the Planning Acts since 1947 are as much part of the legislation as any other provisions. Planning permission granted under section 299 of the 1990 Act would have effect subject to C the restrictions on enforcement contained in sections 294(1) and 296(2)(a). The judge was correct to take the view that he did on that point.

[67] It follows that I am not persuaded that it was an abuse of power for the Home Office to use the NPD procedure in relation to its proposals for the construction of accommodation centres on the former DSDC land at Bicester or, more pertinently in the context of the present proceedings, that D the First Secretary of State had been wrong to entertain a reference under para 23 of that procedure or to proceed to a decision under para 25.

Conclusion

[68] I would dismiss this appeal.

DYSON LJ: E

[69] I agree. I wish to add a few words of my own on the second issue. It is submitted on behalf of the district council that it was an abuse of power for the Secretary of State for the Home Department (SSHD) to use the NPD procedure when he could have applied for planning permission under section 299 of the 1990 Act. Mr Elvin was driven to argue for an abuse of power because, on the face of it, section 299(2) is permissive and F not mandatory "notwithstanding the interest of the Crown in the land in question, an application for any such permission… *may* be made…": my emphasis.

[70] Underpinning Mr Elvin's case on abuse of power is the submission that no proper reason can exist for the Crown taking advantage of a non-statutory procedure that is less protective of the public interest than G the controls of the 1990 Act. But for the reasons explained by Chadwick LJ, the safeguards available where an application is made under section 299 are not materially different from those that are available if the NPD procedure is invoked. The suggested foundation for the abuse-of-power argument is therefore illusory.

[71] Nor is it said that the SSHD used the non-statutory route improperly H in order to achieve some other advantage, or for some other motive. In view of the broad equivalence of safeguards between the statutory and non-statutory procedures, it is difficult to see what improper motive there could be. The reason why the SSHD chose to invoke the NPD procedure in this case, and indeed why he ever does so, is obscure. It is not explained

A in the evidence. Mr Philip Sales said that it was "probably due to the way in which matters developed historically". There was a somewhat faint complaint by Mr Elvin that the SSHD has not advanced reasons for not using the section 299 procedure. But no challenge has been made to the lawfulness of the decision of the First Secretary of State on the ground that the SSHD did not give reasons for adopting the NPD procedure. In my B view, such a challenge could not in any event succeed. There is no basis in law for requiring the Crown to provide reasons for adopting one lawful procedure rather than another lawful procedure.

[72] For these reasons, as well as those given by Chadwick LJ, I would dismiss this appeal.

MUNBY J:

C [73] I agree with both judgments.

Appeal dismissed.

R (on the application of Pridmore and others) *v* Salisbury District Council and another

[2004] EWHC 2511 (Admin)

ADMINISTRATIVE COURT
Newman J
5 November 2004

Planning application – Procedural irregularities – Notices – Certificate B – Application site not wholly owned by applicant – Applicant failing to serve notice on landowner prior to application – Town and Country Planning (General Development) Order 1995 – Certificate B served after application made – Certificate manifestly false – Failure to observe mandatory requirements – Consequences of failures flowing from procedural errors – Whether procedural errors going to jurisdiction of planning authority to determine planning application

In October 2003, D, the interested party, applied for planning permission for residential development of certain land adjoining that of the first and second claimants, PP. D's application certified that he owned all the application site. In fact, a small part of the site belonged to PP; they received a neighbour's notice from the defendant council notifying them of D's application. By late October, the planning officer had knowledge of PP's ownership of part of the land, and that the certificate signed by D was false. The scheme ultimately was not approved. The planning officer recommended that D withdraw it and make a new application, and that he serve an owner's notice on PP and file a certificate B. D did not withdraw the application but amended it, and, at a later date, sent in a certificate B. The planning officer treated the amended application as valid and falling within the requirement that it should be dealt with within the eight-week period prescribed for the handling of applications pursuant to article 20(1) of the Town and Country Planning (General Development Procedure) Order 1995 (the GDPO). No notice of the application was served by D on PP as required by article 6(5)(a) of the GDPO. Planning permission was granted in February 2004. The claimants sought judicial review of that decision.

Held Permission to apply for judicial review was granted and the substantive application was allowed. The decision to grant planning permission to D was quashed.

The planning application was invalid. The planning officer had erred in law in considering there to be a requirement to determine the application within an eight-week period. Consequently, she had wrongly concluded that adequate notice had been given to claimants, as objectors, to consider the amended plan. D should have given prior owner's notice to the claimants, as owners of part of the application site, and D should have completed a certificate in the correct form: see [24] to [26]. The planning officer had failed to exercise the council's discretion in accordance with para 25 of Department of the Environment Circular 15/92 by failing to take into account a number of material considerations: see [31]. The conduct of D disclosed a cavalier disregard for the mandatory requirements in connection with the statutory certificate. The exercise of the court's discretion, so as to preserve the benefit of the planning permission granted to D, would come close to undermining the mandatory scheme of the legislation. It should not be used to save voidable grants of planning permission from being quashed where there has been a deliberate failure to comply with the mandatory requirements of the legislation. The claimants were prejudiced by the procedural errors. *Main* v *Swansea City Council* (1985) 49 P&CR 26 distinguished. The amended plan should have been put before the Highway Authority: see [43] to [44].

Case referred to in the judgment

Main v *Swansea City Council* (1985) 49 P&CR 26; [1985] JPL 558, CA

Permission for judicial review

The claimants, Ian Pridmore, Jo-Anne Pridmore, Terence Dodd and Linda Dodd, sought permission to challenge, by way of judicial review, a decision of the defendants, Salisbury District Council, to grant planning permission on an application by the interested party, Bernard Docking.

Daniel Kolinsky (instructed by Burges Salmon LLP, of Bristol) appeared for the claimants, Ian Pridmore, Jo-Anne Pridmore, Terence Dodd and Linda Dodd.

Simon Pickles (instructed by the solicitor to Salisbury District Council) represented the defendant council.

Bernard Docking, the interested party, did not appear and was not represented.

The following judgment was delivered.

NEWMAN J:

[1] The claimants seek permission to challenge a decision of the defendants, Salisbury District Council (the council), dated 27 February 2004, to grant planning permission to the interested party (Mr Bernard Docking) for the construction of a dwellinghouse at the rear of Maesmor, 38 Firs Road, Firsdown, Salisbury. The first and second claimants live at 36 Firs Road and the third and fourth claimants live at 34 Firs Road. Mr Docking lives at 38 Firs Road. Pitchford J ordered in the application for permission for an oral hearing, with the full hearing to follow should permission be granted.

[2] The claim form contains six grounds of challenge, two of which are not now being pursued. The principal ground that has been pursued raises an important issue as to the consequences that should properly flow from a series of procedural irregularities in connection with the application. It is not disputed that the council's decision is voidable. The argument on this ground has concentrated upon whether the court should exercise its undoubted discretion to quash the decision, having regard to the approach established by a number of cases and, in particular,

Main v *Swansea City Council* (1985) 49 P&CR 26. I propose, having outlined the essential facts, to consider first the ground based upon the irregularities, and I shall return, as necessary, to the other grounds, namely:

(i) a failure to have regard to material considerations relevant to a highway access;

(ii) the giving of misleading guidance on tandem development to the planning committee;

(iii) a misunderstanding, on the part of the officer giving the report to the planning committee, as to the meaning of PPG 3.

Each of (i) to (iii) are contested by the council.

Essential facts

[3] On 14 October 2003, Mr Docking submitted a planning application for a new dwelling at the rear of 38 Firs Road. The plan (the original plan) submitted in support of the application disclosed a proposal for the construction of a four-bedroom bungalow at the rear of his garden, close to the boundary of the garden of the first and second claimants (the Pridmores) at no 36. Number 36 itself occupies what was once part of the rear garden of no 38. The parcel of land was transferred to a Mr and Mrs Collins by a deed of gift on 15 November 1956, when no 38 was owned and occupied by the father of Mrs Collins. By a subsequent conveyance dated 31 August 1964, Mr and Mrs Collins acquired the driveway running from no 36 down to Firs Road. The conveyance included the whole of the land comprised by the driveway to the point where it meets Firs Road.

[4] The original plan showed an intention on the part of Mr Docking to provide a driveway from Firs Road to the proposed bungalow, running down from the immediate area of the bungalow, parallel and immediately adjacent to the Pridmores' drive, along its length to Firs Road. This driveway, according to the plan, was to have a passing bay carved out of land presently owned by Mr Docking, to enable vehicles entering and exiting from the property at the same time to pass without having to reverse on to Firs Road. At Firs Road, the original plan showed that it was possible for a vehicle to enjoy the use of a widened splay before entering Firs Road. Part of the land comprising the splay then belonged, and still belongs, to the Pridmores but, according to the terms of the application for permission, all the land included in the proposed development was certified as being owned by Mr Docking.

[5] On 18 October 2003, the Pridmores received a standard letter from the council notifying them of the application for permission, in their capacity as "occupier of a property in the vicinity of the site" (a neighbour's notice). In common form, the letter gave notice that the plans and other details were available for inspection at the planning office in Salisbury on weekdays between 9am to 5pm and on Friday 9am to 4pm. Because of their working commitments, there was little prospect of the Pridmores gaining access to the plans and details in the near future and so Mr Pridmore went, without delay, to see Mr Docking.

[6] Mr Docking showed Mr Pridmore a copy of the original plan. They also had a discussion in the course of which Mr Docking stated that his architect had mentioned a proposal regarding access to the new property,

which, it appears, the architect had in mind to raise at a later stage, perhaps after permission had been obtained. In essence, the Pridmores could be offered a couple of feet of extra width on their driveway, to be taken from Mr Docking's land, and, in exchange, a turning circle could be created at the top of the drive close to or in the Pridmores' garden in order to service the new property. Mr Pridmore saw the proposal as tantamount to making the driveway a public highway for the benefit of the new property and visitors to the new property.

[7] The original plan, of course, did not reflect this proposal, but it did show a garage for the new property at the southern end of the new plot, close to the proposed boundary with no 38 and situated in an amount of space that was adequate to allow, assuming no other vehicles were parked, a vehicle to turn around before proceeding down the driveway to Firs Road.

[8] Sometime before 22 October, Mrs Wallace, the planning officer for the council, became aware that the Pridmores owned part of the splay and that the certificate signed on behalf of Mr Docking was, consequentially, false. By this time, the Highway Authority (Wiltshire Highway Partnership) had been consulted on the original plan. By a letter dated 30 October 2003, it indicated that it was not raising any objection to the proposal.

[9] On 11 November 2003, Mrs Wallace informed Mr Docking's agent that the proposed dwelling was likely to receive a recommendation for refusal. She suggested that the application should be withdrawn and that Mr Docking could start again. She stated that if the application were not to be withdrawn, since the Pridmores had stated that the applicant did not own all the land within the red line on the plan and that they owned some of it, notice should be served by Mr Docking on the Pridmores as owners of some of the land in the application, and further that an appropriate certificate B should be filed. The reason given for the likely refusal of the application was that the property, as shown on the original plan, was too close to the Pridmores' property. Mrs Wallace suggested that the bungalow should be reduced in size and moved so that the garage, which was presently on the south side of the property, would be situated adjacent to the boundary with the Pridmores, and the bungalow would thereby be farther from the Pridmores' boundary.

[10] At all material times throughout these discussions and thereafter, it would appear that Mrs Wallace treated the application as valid and falling within the requirement that it should be dealt with within eight weeks of its submission: see article 20(1) of the Town and Country Planning (General Development Procedure) Order 1995 SI 1995/419 (the GDPO). As a result, the application was to be considered at a planning committee meeting on 16 December 2003.

[11] The application was not withdrawn by Mr Docking; instead, an amended plan was submitted on 27 November 2003, being a little under three weeks before the proposed planning committee meeting. No form of certificate B was delivered with the amended plan. It followed that

there was no confirmation that an owner's notice had been served on A the Pridmores. Notice of any application for planning permission must be served on an owner of land included in the application before the application is made: see article 6(5)(a) of the GDPO. Since Mr Docking had not made a fresh application, this requirement could not be met in connection with the application, which he had chosen to amend. Mrs Wallace did not receive any form of notice that the Pridmores had B been served as owners (albeit out of time), and did not receive a form of certificate B until 10 December 2003, being six days before the proposed meeting. Despite the absence of any form of certificate B, Mrs Wallace nevertheless informed all objectors (including the Pridmores) by a letter dated 9 December 2003 that the application was to be considered on 16 December 2003. The Pridmores had already received a letter notifying C them, as neighbours, of the amended application on 3 or 4 December. The amended plans were not submitted to the Highway Authority at any time before the committee meeting on 16 December 2003.

[12] Certificates and notices are required to be in the form prescribed by the GDPO. Certificate B, when submitted, purported to verify that the Pridmores had received notice of the application as D owners on 2 December 2003. It was manifest on the face of the certificate that the notice required by the Town and Country Planning Act 1990 (T&CPA) and the GDPO had not been given. Mr Docking had elected not to submit a fresh application, and the failure on his part to serve an owner's notice before the application had been made could not be fulfilled. Nor, having regard to the date set for the hearing of the application, and E disregarding the falsity of the certificate, could 21 days elapse from the date when the Pridmores had notice of the amended application: see article 20(5)(b) of the GDPO. It is to be remembered that the Pridmores had received due notice of the application, as neighbours, and, as we shall see, the requirements for notice to objectors of any amendment to an application, being discretionary, are less strict: see para 25 of the F Department for the Environment (DOE) Circular 15/92.

[13] However, it has to be pointed out that at no time did the Pridmores, according to their evidence, receive any notification as owners (even late notice) that a planning application had been made that affected land owned by them. There is no evidence from Mr Docking or his agent on the issue, but for the terms of the certificate itself, that the Pridmores G were given notice on 2 December. On the evidence, the form of certificate B lodged by Mr Docking must be regarded as false and the Pridmores' evidence, if wrong, should have been answered. It should be remembered that section 65(6) of the T&CPA makes it an offence to knowingly issue a false or misleading certificate.

[14] The planning officer's report to the committee was drafted on H 9 December 2003, being the day before the receipt by the council of the form of certificate B. On 16 December 2003, the planning committee considered the application, having considered the report and the matters raised before them, and resolved to grant planning permission. Planning permission was issued on 27 February 2004.

Statutory requirements in relation to notification

[15] The regime commences with the T&CPA and the details are laid down by the GDPO. Section 65(2) of the T&CPA requires that provision should be made by development order for the notification of owners of land upon which planning permission is sought. Section 65(5) provides:

> A local planning authority shall not entertain an application for planning permission unless any requirements imposed by virtue of this section have been satisfied.

I have already drawn attention to the penalty for a knowingly committed breach.

[16] Effect has been given to section 65 of the T&CPA by the GDPO. Article 6(1) of the GDPO prescribes that:

> Subject to paragraph (2), an applicant for planning permission shall give requisite notice of the application to any person (other than the applicant) who on the prescribed date is an owner of the land to which the application relates...
>
> (a) by serving the notice on every such person whose name and address is known to him...

By article 6(5)(a), the "prescribed date" is 21 days before the date of the application. It follows that, once an application has been submitted without due notice to an owner, the requirement cannot be fulfilled. The consequences of non-fulfilment are that the application should not be entertained by the local planning authority (see section 65(5)) and the application should not be determined (see later under article 20(5) of the GDPO). On this occasion, the local authority knew of two false certificates, knew that notice had not been given and that it was possible for it to be given and proceeded to entertain and determine the application.

[17] Article 7 of the GDPO provides for certificates to be in a prescribed form. It provides:

> Where an application for planning permission is made, the applicant shall certify, in the appropriate form prescribed in Part 2 of schedule 2 to this Order or in a form substantially to the like effect, that the requirements of article 6 have been satisfied...

It can be seen from the detail of the above provisions that the clear intention of the statutory scheme is that an owner of land included in the proposed development should be given prior notice of an intention to submit a planning application, and that the obligation to give such notice falls on the applicant for permission.

[18] Article 8 of the GDPO imposes an obligation upon the local planning authority to notify adjoining owners or occupiers. The council discharged this duty.

[19] Article 19 provides:

> (1) A local planning authority shall, in determining an application for planning permission, take into account any representations made, where any notice of the application has been –

> (a) given by site display under article 6 or 8, within 21 days beginning with the date when the notice was first displayed by site display;
> (b) served on
> (i) an owner of the land or a tenant of an agricultural holding under article 6, or
> (ii) an adjoining owner or occupier under article 8,
>
> within 21 days beginning with the date when the notice was served on that person, provided that the representations are made by any person who satisfies them that he is such an owner, tenant or occupier; ...

It follows that the right to receive a notice gives rise to a requirement, on the part of the recipient, to make representations within 21 days in order, in turn, to impose an obligation on the local planning authority to take the representations into account.

[20] Article 20 of the GDPO is headed "Time periods for decision". In its material part it provides:

> 20.– (1) Subject to paragraph 5, where a valid application under article 4 or regulation 3 of the 1988 Regulations (applications for planning permission) has been received by a local planning authority, they shall within the period specified in paragraph (2) give the applicant notice of their decision or determination or notice that the application has been referred to the Secretary of State.

The relevant period, according to subsection (2), is a period of eight weeks beginning with the date when the application was received by the local planning authority. Subsection (3) states, as material:

> For the purposes of this article, the date when the application was received shall be taken to be the date when each of the following events has occurred –
> (a) the application form or application in writing has been lodged with the authority mentioned in article 5(1);
> (b) any certificate or documents required by the Act or this Order has been lodged with the authority;
> (c) ...

Article 20(5) provides:

> Subject to paragraph (6), a local planning authority shall not determine an application for planning permission, where any notice of the application has been
> (a) given by site display...
> (b) served on... an owner of the land... or to an adjoining owner or occupier under article 8
>
> before the end of the period of 21 days beginning with the date when the notice was served on that person;...

[21] For completeness, it should be noted that there is no statutory obligation on local planning authorities to publicise changes to applications once they are accepted as valid. It is at the discretion of the authority to decide whether further notice is desirable: see para 25 of DOE Circular 15/92.

[22] In summary, the procedural position is as follows:

A (1) Mr Docking failed to notify the Pridmores of the planning application as he was required to do pursuant to section 65(5) of the T&CPA and article 6 of the GDPO.

(2) Mr Docking (or his agent acting on his behalf) completed a false certificate of ownership that wrongly certified that he owned the whole of the land covered by the application.

B (3) Mr Docking (or his agent acting on his behalf) thereafter completed a certificate that falsely stated that the application had, on 2 December 2003, been served on the Pridmores as owners of part of the land included in the application. On the evidence before me, no such form of notice was served on 2 December 2003, or at any other time.

(4) Since the application was dated 14 October 2003, it was manifest
C that, even if notice had been given on 2 December 2003, prior notice of the application had not been given. It has not been suggested that prior notice of an amended application, where the original application has not been the subject of prior notice, satisfies article 6 of the GDPO.

(5) Between 22 October (at the latest) and thereafter, the council entertained the application knowing that false certificates had
D been lodged.

(6) The application for permission was determined on 16 December 2003. Even if the false revised certificate had been regarded as valid, article 20(5)(b) stipulates that the planning application cannot be determined within 21 days of the service of the article 6 notice upon the owner.

E (7) The Pridmores received notice of the original and amended application from the council, given to them in their capacity as neighbours. Notice of the amended application was given less than two weeks before the date on which the matter was determined.

[23] It is submitted, on behalf of the council, that Mrs Wallace entertained the application and submitted it for determination because
F she believed that the council were under an obligation to determine the planning application within eight weeks of its receipt by the council, namely on 14 October 2003. The evidence does not disclose that any consideration was given to the terms of article 20 of the GDPO. The eight-week rule applies to a "valid application", and time does not start to run until any certificate required by the GDPO has been received:
G see article 20(3).

[24] One must be careful in this case not to conflate the separate issues concerning the invalidity of the application for permission and the considerations relevant to the application to quash the consequent grant of planning permission. Having regard to the language of section 65(5) and articles 6 and 20, I have no hesitation in concluding that the requirements
H imposed upon an applicant are mandatory. A dishonest failure to adhere to them is a criminal offence. The application was invalid. But this conclusion still leaves open the answer on the issue as to whether the grant of permission should be quashed. It follows that Mrs Wallace erred in law in considering that there was a requirement to determine the application within eight weeks. The immediate consequence of the error was that she

concluded that adequate notice had been accorded to the Pridmores, as objectors, for them to consider the amended plan.

[25] Further, in communicating with the agent for Mr Docking on 11 November, she did not appear to realise that it was open to the council to refuse to accept the application on the ground that the requisite owner's notice had not been served.

[26] In my judgment, she should have required a fresh application to be made, with prior notice being given to the Pridmores and the completion of a certificate in the correct form. No complaint could have been made had she refused to accept the application. In para 8 of her witness statement, she states:

> I asked that if they wished to continue with the application then as Mr and Mrs Pridmore's representations indicated that the Applicant did not own all the land within the red line, I asked for notice to be served on the neighbour at No 36 and for an appropriate Certificate B to be filed with the application.

It is difficult to see what it was hoped would be achieved by this if, by "the application", she meant the existing one, since an appropriate certificate B could not be filed in connection with the existing application. If she meant to refer to the application continuing by means of a fresh application being lodged, her subsequent behaviour is difficult to understand, for she accepted an amended application.

[27] It is not clear from para 8 whether Mrs Wallace realised on 8 November that an amended plan might be lodged. Paragraph 9 is equally unilluminating on the issue. She merely states:

> However, the Applicant was unwilling to withdraw his application and, on 27 November, his agent submitted amended plans.

No doubt because the amendments reflected the advice she gave on 11 November, she does not appear to have considered the changes as they might have affected vehicles turning in the garage area to leave the property. She states, in para 9:

> I returned the amended application form to the agent because details regarding the Certificate had not been completed correctly. I asked for the correct completion of the Certificate B and confirmation of when notice had been served on the neighbour at No 36.

The nature of the error that caused her to send the amended application back is not disclosed. It follows that Mr Docking, by his agent, committed these errors in connection with the requisite certificates. The suggestion by Mrs Wallace that there could be a correct completion of certificate B could only arise if she thought that the Pridmores had been notified prior to 14 October (the date of the application). Since the error originally made was to ignore their ownership and certify that the applicant was the owner of all the land, it is difficult to see what Mrs Wallace had in mind.

[28] Whatever she had in mind on or about 27 November, it is difficult to see why she was satisfied with the certificate B that she did receive, eventually, on 10 December. In the meantime, she had continued to process

the application and sent out notices in connection with the amended plans and the date for the committee hearing.

[29] Although there is no reason to believe that she knew that no notification had been given to the Pridmores by the applicant or his agent on 2 December 2003, had she considered the terms of the certificate B, she must have realised that it was false, although it purported to be in the prescribed form, because she knew no notice prior to the application had been given.

[30] Further, she gave no effect to the form of notice by following article 20(5)(b) of the GDPO, which requires 21 days to expire before the planning application is determined.

[31] In the light of the above, I have concluded that, in so far as she exercised the council's discretion in accordance with para 25 of DOE Circular 15/92, she erred because she failed to take account of a number of material considerations. To the matters appearing above must be added the failure to consider whether the amended plans should be submitted to the Highway Authority, and the contents of [41] to [44] below.

[32] Having stated the character of the irregularities, I turn now to the issue of discretion, commencing with the law.

Law on voidable decisions
Main v Swansea City Council
Facts

[33] An application was made for outline planning permission for a residential development of certain land, and the certificate stated that requisite notice of the application had been given to all other owners of the land. A small, but not *de minimis*, part of the land was owned by another person, who was not specified and whose identity was and subsequently remained unknown. The scheme ultimately approved did not involve development of the land owned by the unspecified person. Outline planning permission had been granted in January 1977, and approval of the reserved matters was granted on 24 April 1980. Mr Main was granted permission to apply for judicial review on 12 June 1980. The inference from the material before the court led it to conclude that the original application was unaccompanied by a certificate of any kind as required by the relevant section of the Act and that, when that was observed, it was returned by the local planning authority with a request that it should be re-submitted with an appropriate certificate.

[34] The court concluded that, at the time when the council considered and determined or entertained the outline application, it was accompanied by a certificate that, in form, was, on its face, as required by the relevant section of the Act. But, as the court observed, at p31:

> That certificate, however, was not the certificate that, on the facts, was required by section 27. The outline planning application related to land that, although almost wholly owned by the council, was owned as to a small, but not *de minimis*, part, by some person or persons whose identity was at the time, and still remains, unknown.

[35] Accordingly, the court concluded that the certificate should have A
stated, and included a statement to the effect, that the applicants had taken steps (and specifying them) as were reasonably open to them to ascertain the names and addresses of the owners of the part of the land not owned by the council, but had been unable to do so. It should have also stated that requisite notice had been given by being published in a newspaper circulated in the district. The court therefore concluded: B

> (a) the certificate was inaccurate in what it did state, for notice had not been given to all the owners; and (b) that the newspaper publication designed to ensure that the application should come to the notice of unidentifiable owners had not been effected.

[36] It had also been contended in the case that the certificate was known C
to be false both by the council and by the agents who had submitted it. This contention had been rejected by the trial judge, and it was rejected by the Court of Appeal. The Court of Appeal stated:

> There is, however, no evidence to establish that, either when the certificate was submitted or when the outline permission was granted, it was known to the council or the applicants or their agents that it was false. D

[37] After detailed consideration of the principles in connection with jurisdictional and procedural error, and the circumstances in which the former will arise so as to give rise to a complete nullity as opposed to a decision being vitiated and voidable, it held:

(i) a factual error in a certificate under section 27 of the Town and E
Country Planning Act 1971 might be no more than an irregularity that did not go to the jurisdiction of the planning authority to entertain the application for planning permission, notwithstanding that the provisions included the words "shall not entertain";

(ii) it rejected a submission that a factual error that was not so gross as to make the certificate no certificate at all was necessarily a mere irregularity F
in respect of which an applicant was not entitled to relief;

(iii) the defects in the certificate had been sufficient to entitle the court to strike down the subsequent grant of outline planning permission in certain circumstances, but that it was not a complete nullity and the matter was one for the discretion of the court;

(iv) since, throughout the period between the grant of outline G
planning permission and the approval of reserved matters, the applicant had not objected to the outline planning permission, and the Secretary of State had known of the position with regard to article 11 for a long time and not sought relief, and since the scheme approved did not involve development of the land owned by the unidentified person, it was too late, at any rate at the suit of the applicant, to quash the H
grant of outline planning permission and thereby the approval of reserved matters.

[38] Mr Simon Pickles, for the council, submitted as follows:

(i) In this instance, there is no allegation of dishonesty against his client nor any allegation of bad faith.

A (ii) Although Mr and Mrs Pridmore suggested that they were provided with inadequate time within which to consider an objection, which they would have advanced to the amended plans in connection with the availability of turning space, in truth they had had sufficient time to respond. He submitted that they had effectively been engaged in the planning process from an early stage. They had viewed the B revised drawings.

(iii) Although the procedural irregularity vis-à-vis the Pridmores as owners could not be corrected, the overlap between their status as owners and their status as a neighbour meant that they had been involved in the consultation process. They had made representations and they had been considered.

C (iv) The decision should not be quashed unless at least there was a possibility of a different outcome had the regulatory breach not occurred.

(v) The council had acted quite properly in bringing the revised drawings to the Pridmores' attention and, although they had not received 21 days for response, they had received an adequate period of time that was reasonable.

D (vi) He submitted that the Pridmores had raised their complaint very late in these proceedings. They did not request further time at either consultation stage. He submitted that the case for prejudice that they are now seeking to make out had only been raised at a late stage in these proceedings. He made no submission in connection with the conduct of Mr Docking or his agent.

E [39] In my judgment, the conduct of Mr Docking and/or his agent on his behalf disclose a cavalier disregard for the mandatory requirements in connection with a statutory certificate. It is one thing to fail to give notice to an unidentified owner of part of the land, as in the case of *Main*, but it is quite another to certify that prior notice of an application has been given, when it is known that no such notice has been given. More than that, as F I am satisfied on the evidence, it is yet worse to certify that notice has been given on a stated date (2 December) when no notice has been given on that day or at all.

[40] I have concluded that for the court, in these circumstances, to exercise its discretion, so as to preserve the benefit of the planning permission granted to Mr Docking, would come close to undermining the G mandatory scheme of the legislation. Nor can I see that the law as stated in *Main* anticipates saving voidable grants of planning permission from being quashed where there has been a deliberate failure to comply with the mandatory requirements of the scheme of the legislation to which the local planning authority, acting in good faith, have been party and where, operating under an error of law, they have gone on to press, unnecessarily, H for prompt and timeous determination. In *Main*, the applicant for relief was not the owner affected by want of notice and the allegation of acting with knowledge of the false certificate was rejected.

[41] More than that, I reject the suggestion that it is clear that no prejudice has occurred. It is not a convincing answer to the Pridmores' complaint, which is that more time was required to consider the amended plan, for

the council to assert that the matter now raised in these proceedings should have been raised earlier and/or that they should have asked for more time at the time they saw the amended plan. The suggestion misses the substance of the complaint. It is because they had too short a period of time to consider the amended plan that the Pridmores now suggest that they failed to realise that points on the available turning space could have been made by reference to the amended plan. For that reason, the issue has been raised in these proceedings.

[42] As I pointed out in the course of argument, it is by no means clear how much space is available, according to the amended plan, to enable vehicles to turn. The suggestion made by Mr Docking to Mr Pridmore that the architect had in mind to suggest negotiations for the creation of a turning area points to the existence of a tenable view that, even with the plan as submitted, turning was considered likely to give rise to difficulty.

[43] Nor am I satisfied that sufficient attention was given by Mrs Wallace to a risk that the amended plan might cause vehicles to reverse into Firs Road. Although consideration was given to the passing bay as obviating the need for a vehicle that was entering the property to reverse into Firs Road, in my judgment, the terms of the report to the committee could have been understood as stating that no risk of reversing otherwise existed. In the light of the evidence before the court, these matters require investigation and consideration.

Highway Authority

[44] I have seen no explanation for the amended plan not being put before the Highway Authority. I assume the failure is to be explained by the desire to adhere to the 16 December date for the decision. I have seen evidence relating to what the Highway Authority might have considered had the amended plan been submitted to them. From the statement of Mrs Wallace, which records this evidence, fresh consideration was given only to the significance that the Highway Authority would have attached to the fact that the splay at Firs Road was partly owned by the Pridmores. I regard this as an indication that the point being made by the Pridmores in their representations to the planning committee was not adequately considered by Mrs Wallace. The fact of ownership carried with it a related point in connection with a right of way over the land in the splay. Whereas Mr Docking enjoys a right of way that would enable him, when turning left into Firs Road, to angle his car away from the bend to the driver's right, unless the Pridmores accord the owners of the new property an equivalent right of way, that facility will not be available and the emerging vehicles will be closer to the bend and will not be able to be angled for the manoeuvre. On the evidence, the Pridmores are not prepared to grant a right of way. At the least, this seems to require investigation and consideration by the Highway Authority along with the point made in connection with the turning facilities for the property.

[45] For the reasons that I have stated, and in the exercise of my discretion, I have concluded that the planning permission granted in this case must be quashed, on this ground alone.

[46] Having regard to my conclusion, I propose to state my conclusions on the remaining grounds briefly.

Failure to have regard to material considerations relevant to highway access

[47] I have said enough in [43] to [46] above to show that I have concluded that there was a failure to have regard to relevant considerations. I have little doubt that Mrs Wallace permitted her erroneous view of the eight-week requirement to influence her decisions.

Misleading guidance on tandem development

[48] The guidance is said to have been misleading because it failed to distinguish between a back garden development (being the first in the garden in question) and two levels or occasions of development within a back garden area (being the second new house in the original garden area). I see no reason to confine the concept of tandem development to one development per garden. Much will depend upon the size of the garden in the first place. It is the density of the development that matters, and that can be ascertained from the plans and the site and the housing in the area. It must have been obvious to an informed planning committee that the Pridmores' property had been built in the garden of no 38.

Misunderstanding of PPG 3

[49] Mrs Wallace's report stated:

> tandem development is already present in the locality and it may be difficult in principle especially in view of the present requirement to increase densities in existing urban areas, to refuse a new dwelling in this position.

It is submitted that it was wrong to suggest that national planning policy (PPG 3) "required" an increase in "densities". Further, it was submitted that the suggestion amounted to advice that the policies in the adopted development plan, which was resistant to tandem development, had been superseded.

In my judgment, the submission ignores the impact of the next paragraph of the report to the following effect:

> Although the proposed dwelling is located in a backland position, it is considered that it is an acceptable form of development on account of the design which is single storey, low impact and has no window openings above the ground floor that are likely to overlook neighbouring properties. The proposed dwelling would be within a plot smaller than some in the local area but very similar to those in Ilynton Drive and Juniper Drive.

This paragraph follows the criteria set out by the adopted development plan. Had this paragraph followed the opening words of the previous paragraph instead of the reference to "the present requirement to increase densities", the position would have been clearer.

I am satisfied that, in the context of the report, it was a form of shorthand to refer to national policy as a "requirement". It should be remembered that national policy does impose some procedures to be followed, for example, a search sequence and maximisation of previously developed land and the more efficient use of land. In a loose sense, these are goals

that the local planning authority have to have in mind and they are, by A
their nature, going to have the consequence, when applied, of increasing density in urban areas.

I regard the criticism of the report to be more as to form than substance.

[50] Permission to apply for judicial review is granted and the application is granted. The decision to grant planning permission to the
interested party is quashed. B

Application granted.

R (on the application of Blewett) *v* Derbyshire Waste Ltd

[2004] EWCA Civ 1508

COURT OF APPEAL
Auld, Buxton and Laws LJJ
11 November 2004

Landfill – Waste – Waste Framework Directive 75/442/EEC and Landfill Directive 1991/31/EEC – Relevance of article 8(b) of Landfill Directive in domestic law – Planning application for waste disposal by landfill – *Waste Strategy 2000* – Best practicable environmental option (BPEO) – Whether BPEO analysis mandatory in waste-planning development control decisions – Whether authority failing to carry out BPEO analysis – Whether authority's decision unlawful

In 2002, the local planning authority granted planning permission to the appellant, D Ltd, in respect of a site for waste disposal by landfill. The respondent, B, applied by way of judicial review to quash that decision. Sullivan J allowed his claim, concluding that the planning committee should not have granted planning permission without first satisfying themselves, pursuant to article 8(b) of the Landfill Directive, that the proposal was "in line with" the best practical environmental option (BPEO). The BPEO, which was incorporated in a national plan, *Waste Strategy 2000*, established a hierarchy of the desirability of methods of confronting waste. The authority had failed to prepare a waste-management plan in accordance with the Waste Framework Directive, although they had a non-statutory waste-management strategy that reflected national guidance in the form of the BPEO. The judge held that even if the authority had not been so obliged, their consideration of the BPEO had been so inadequate as to render their grant of planning permission unlawful. Once the authority had concluded that a capacity existed, both locally and countywide, up to 2011, the question as to whether the particular application site would be the BPEO for meeting a shortage of landfill space in the relevant area had not been addressed in terms of the three key considerations: namely (i) the waste hierarchy; (ii) the proximity principle; and (iii) self-sufficiency in managing waste. The appellant and the Secretary of State appealed that decision.

Held The appeal was dismissed.

Article 8(b) of the Landfill Directive was part of domestic law: see [62]. Alternatively, there was a strong common-sense imperative to construe the Waste Management Licensing Regulations 1994 and any waste-management plan made or amended pursuant to the obligation under article 7 of the Waste Framework Directive that they implement, as forming part of the same mechanisms for attaining the objectives in articles 3, 4 and 5 of that directive: see [63]. Article 8(b) of the Landfill A Directive does not make it a precondition of the grant of planning permission for a landfill proposal that that proposal should satisfy the policies relating to the BPEO as incorporated in a waste-management plan made pursuant to article 7 of the Waste Framework Directive, such as *Waste Strategy 2000*. However, the council's consideration of the BPEO had been flawed in any event. *Waste Strategy 2000* and the BPEO are B required to be taken into account in both plan-making and development control. An authority cannot rely upon their own failure to provide a waste-management plan as a reason for not carrying out a full BPEO analysis: see [111]. Sullivan J had been entitled to conclude that the authority had not demonstrated that they had given sufficient importance or weight to the BPEO policies: see [113]. It was not appropriate for C the court to take a different view from Sullivan J's as to the exercise of discretion. To do so would amount to a consideration of the planning merits. It was not possible to say that the council would have taken the same decision had it sensed the greater importance that should be given to the BPEO hierarchy: see [119].

D

Cases referred to in the judgment

AvestaPolarit Chrome Oy v *Lapin Ympäristökeskus* C-114/07 [2003] ECR I-8725

Belgium v *Traen* C-372/85 [1987] ECR 2141, ECJ

Berkeley v *Secretary of State for the Environment, Transport and the Regions (No 1)* [2001] 2 AC 603; [2000] 3 WLR 420; [2000] 3 All ER 897; (2001) 81 P&CR 35; [2000] 3 PLR 111; [2001] JPL 58, HL

Comitato di Coordinamento per la Difesa della Cava v *Regione Lombardia* C-236/92 E [1994] Env LR 281

Commission of the European Communities v *France* C-292/99 [2002] ECR I-4097, ECJ

Commission of the European Communities v *Greece* C-387/97 [2000] ECR I-5047, ECJ

Commission of the European Communities v *Italy* C-365/97 [1999] ECR I-7773, ECJ

Commission of the European Communities v *United Kingdom* C-35/00 [2002] ECR I-953, ECJ

F

Commune de Braine-le-Château and Tillieut v *Région Wallonne* C-53/02 and C-217/02, unreported 1 April 2004, ECJ

G v *G* [1985] 1 WLR 647; [1985] 2 All ER 225, HL

Levy v *Environment Agency* [2002] EWHC 1663 (Admin), [2003] Env LR 11

Marks & Spencer plc v *Commissioners of Customs & Excise* C-62/00 [2003] QB 866; [2003] 2 WLR 665; [2002] STC 1036

Marleasing SA v *La Comercial Internacional de Alimentación SA* C-106/89 [1990] ECR G I-4135, ECJ

R (on the application of Blewett) v *Derbyshire Waste Ltd* [2003] EWHC 2775 (Admin); [2004] JPL 751

R (on the application of Burkett) v *Hammersmith and Fulham London Borough Council; sub nom R* v *Hammersmith and Fulham London Borough Council, ex parte Burkett; Burkett, Re* [2002] UKHL 23; [2002] 1 WLR 1593; [2002] 3 All ER 97; [2002] 2 PLR 90

H

R (on the application of Thornby Farms Ltd) v *Daventry District Council; R (on the application of Murray)* v *Derbyshire County Council* [2002] EWCA Civ 31; [2003] QB 503; [2002] 3 WLR 875; [2002] 2 PLR 21; [2002] JPL 937

R v *Leicestershire County Council, ex parte Blackfordby and Boothorpe Action Group Ltd* [2000] JPL 1266; [2001] Env LR 2, QB

Appeal against a decision of Sullivan J

This was an appeal by the appellant, Derbyshire Waste Ltd, against a decision of Sullivan J on a claim by the respondent, John Blewett, for judicial review, quashing a planning permission granted by Derbyshire County Council to the appellant for landfill waste disposal.

Christopher Katkowski QC and John Barrett (instructed by Walker Morris, of Leeds) appeared for the appellant, Derbyshire Waste Ltd.

David Wolfe (instructed by Public Interest Lawyers, of Birmingham) represented the respondent, John Blewett.

James Maurici (instructed by the solicitor to the Department for Environment, Food and Rural Affairs) represented the Secretary of State, the intervener.

The following judgments were delivered.

AULD LJ:

[1] This is an appeal by Derbyshire Waste Ltd (Derbyshire Waste) from the decision of Sullivan J on 7 November 2003 ([2003] EWHC 2775 (Admin)), on a claim by Mr John Blewett for judicial review, quashing, on one of three grounds, a planning permission given by Derbyshire County Council (the council) on 23 December 2002 to Derbyshire Waste to use land at the former Glapwell Colliery, Sutton Scarsdale, near Bolsover in north-east Derbyshire, for "land reclamation by waste disposal with restoration to agricultural, woodland, grassland and nature conservation".

[2] The council have taken no part in the appeal. However, the court has granted the Secretary of State permission to intervene in the matter, which he does by counsel. The court has also permitted Friends of the Earth to file written evidence, which it has done in the form of a helpful witness statement of Dr Anna Watson, who is employed by it, as a waste and resources campaigner.

[3] The main issue in the appeal is whether a waste-planning authority, in determining a planning application for waste disposal by landfill, are obliged by the provisions of two European Community directives, Council Directive 75/442/EEC on waste, which, in its amended form[1] is known as the Waste Framework Directive, and the Landfill Directive 1991/31/EC concerning the treatment of waste, to satisfy themselves before granting permission that the proposal complies with the decision-making policies and methodology for the choice of the method of waste management, namely the "Best Practicable Environmental Option" (BPEO), incorporated in a national plan, *Waste Strategy 2000*, required by the former of the directives. The BPEO, put shortly, establishes for waste-planning decision makers a hierarchy of desirability of methods of confronting waste, starting at the top with reduction of its generation and use of resources, and ending at the bottom with disposal, for example by landfill. All other things being equal, the BPEO methodology is that landfill should be considered last because it is the least desirable environmental option.

[1] As amended by Council Directive 91/692/EEC and Commission Decision 96/350/EEC

[4] Sullivan J held that the council, in making their planning decision, A
were obliged to comply with the BPEO methodology, but that even if they
had not been so obliged, their consideration of the BPEO, in this case,
had been so inadequate as to render their grant of planning permission
unlawful. Derbyshire Waste and the Secretary of State maintain that the
judge was wrong to hold that the council had such an obligation, and
Derbyshire Waste alone argues that he wrongly held, on the facts, that the B
council's consideration of the BPEO had been so inadequate as to render
their decision unlawful. Mr Blewettt and Friends of the Earth maintain
that he was right to hold that the council had such an obligation, the latter
describing the BPEO process "as a mandatory process in waste planning
decisions". And Mr Blewett supports the judge's decision that, in any
event, the council had given inadequate consideration to the BPEO. The C
Secretary of State took no stand on that alternative issue of fact.

Facts

[5] I adopt with gratitude Sullivan J's summary of the facts in his judgment.

[6] In the mid-1970s, Glapwell Colliery closed, leaving two spoil tips.
In the early 1980s, planning permission was granted for a reclamation D
scheme, which involved tip-washing, opencast mining of shallow
seams under the spoil tips and the replacement of the opencast mine
spoil and washed deep mine spoil into a landscaped profile. Voids had
been created within the re-profiled spoil tips as part of the reclamation
works to facilitate landfills. One of them, Smith's Void, or as it came to be
known, Glapwell 3, was to be re-profiled as part of these operations, but E
the contractor employed to carry out the coal recovery scheme went into
receivership, leaving the scheme incomplete.

[7] Glapwell 1 was the first of the voids to be filled. Over a five-year
period between 1983 and 1988, it received and accommodated some
750,000 cubic metres of waste. Planning permission was granted in 1984
for the filling in of two further voids, Glapwell 2 and 3. Waste disposal in F
Glapwell 2 began in 1988, and finished in November 2002, after planning
permission had been granted in 1995 for additional tipping. No tipping
took place in Glapwell 3 pursuant to the 1984 permission, but that planning
permission expired in December 2003 (operations were limited to a period
of 15 years from the start of tipping). The 1984 planning permission had
envisaged that Glapwell 3 would have a capacity of around 1m cubic G
metres. The present proposal for it involves tipping around 850,000 cubic
metres of domestic, industrial, commercial and inert waste over a period of
four years, with the overall operational programme, including restoration
to agriculture etc, taking six years.

[8] The application site covers around 9ha and is located within
1km of the villages of Glapwell, Palterton, Bramley Vale and Doe Lea. H
Mr Blewett lives in Bramley Vale. The nearest boundary of Glapwell 3 is
approximately 800m from his home and those of the existing tipped voids,
Glapwell 1 and 2, around 200m away.

[9] Mr Blewett is registered as a disabled person, suffering from
chronic bronchitis, asthma and angina. He contended that the dust and

A smells generated by the land-filling operations at Glapwell 1 and 2 had exacerbated those conditions. He also complained of other incidents of the land-filling operations, including the attraction to the area of rats and seagulls, the former killing some of his pet pigeons and the latter causing pollution by their droppings. He opposed Derbyshire Waste's application for the grant of planning permission for Glapwell 3, and made B representations to the council in his own right and as a member of the Stop the Landfill Group.

[10] The council are both the waste-planning authority and the waste- disposal authority for their county area. In their latter capacity, and pursuant to arrangements made under section 30 of the Environmental Protection Act 1990 (EPA), they had formed Derbyshire Waste to dispose C of the county's waste. Derbyshire Waste is owned as to 20% by the council, and disposes of the county's waste under a long-term contract with them.

[11] The development proposed by Derbyshire Waste for Glapwell 3, the subject of its planning application, was a "Schedule 2" development as defined in the Town and Country Planning (Environmental Impact Assessment)(England and Wales) Regulations 1999 (the 1999 Regulations)[1]. D Those regulations required an environmental statement if the proposed development was likely to have significant effects upon the environment by virtue of factors such as its nature, size or location. Derbyshire Waste duly included such a statement with its application, which it made on 8 February 2001, justifying the proposal on two main grounds, namely to reclaim a despoiled site and to facilitate the disposal of wastes arising in E the area.

[12] More than a year later, on 11 March 2002, the council's regulatory planning and control committee (the council's committee) considered the application, assisted by a 55-page report from the council's director of environmental services as to its merits (the report). At that time, although the UK had complied with part of its obligation under article 7 of the F Waste Framework Directive, by producing a national waste plan in the form of *Waste Strategy 2000*, it was in breach of its Treaty obligations in the case of Derbyshire Council, which, as the local waste-planning authority, had not produced their own local waste-management plan: see [32] below. In the report, the director recommended the grant of planning permission, but subject to no less than 53 conditions. In so recommending, the director G summarised his reasons in the following way: (1) the proposal would be unlikely to cause significant harm to the amenities of nearby residents; (2) there would be no shortage of landfill space in the county, considered as a whole, before 2010 if reduced waste production, about which there was some uncertainty, and landfill targets were achieved; (3) the proposal was not so large as to create a substantial excess capacity; (4) it would, H in any event, assist in meeting disposal needs in the south-east of the county where an imminent shortfall of landfill space was expected; and (5) the proposal was probably the only satisfactory way of achieving a

[1] SI 1999/293

sympathetic and worthwhile restoration of the site. He concluded that A the proposal accorded with the relevant development plan policies, national policy and regional policy and guidance, specifically mentioning, as part of national policy, *Waste Strategy 2000,* and its incorporation of the BPEO. And he reported that he had taken into account other material considerations.

[13] However, on the morning of the very day upon which the B council's committee were considering the matter, the Secretary of State issued a direction, under article 14 of the General Development Order, preventing the council from determining the application, with a view to calling it in for his decision. The committee resolved that, had they been permitted to determine the application, they would have granted it, as recommended in the report, subject only to a minor amendment to one of C the recommended conditions.

[14] On 22 April 2002, Mr Blewett applied for permission to claim judicial review on a number of grounds, including one that the committee should not have resolved to grant planning permission without first satisfying themselves, pursuant to article 8(b) of the Landfill Directive, that the proposal was "in line with" the government's waste-management plan D for England and Wales, namely *Waste Strategy 2000,* in particular, that the proposed landfill was the BPEO for the waste stream(s) in question. That strategic plan had been made and submitted to parliament in May 2000, pursuant to the UK's obligations under the Waste Framework Directive, as implemented into our domestic law by section 44A of, and Schedule 2A to, the EPA, and, in relation to the diversion of waste from landfill, by article 5 E of the Landfill Directive.

[15] Mr Blewett made his challenge at that stage as a precaution, since it was then uncertain whether the three-month period prescribed by CPR 54.5(1)(b) ran from the date of a resolution to grant planning permission or from the date of the permission itself. In the event, the Secretary of State decided not to call in the application. However, F Mr Blewett's claim for judicial review had by then been overtaken by the House of Lords' decision in *R (on the application of Burkett)* v *Hammersmith and Fulham London Borough Council* [2002] UKHL 23; [2002] 1 WLR 1593[1] that the three-month time limit ran from the date upon which planning permission had actually been granted, and not from the date of an earlier resolution to grant it. Although, as a result, Mr Blewett's application G for permission to claim judicial review had been premature, the court adjourned it to enable him to challenge the grant of permission in due course, if so advised.

[16] On 4 November 2002, the council's committee reconsidered the application. On this occasion, in addition to the report put before them at their meeting of 11 March 2002, they were provided with a report prepared H jointly by the county secretary and the director of environmental services responding to the contentions of Mr Blewett in his judicial review challenge

[1] [2002] 1 PLR 90

A (the joint report). They confirmed the recommendation in the earlier report, and the committee again resolved to grant permission, which was granted on 23 December 2002.

[17] With permission from Collins J, on 29 April 2003, Mr Blewett sought judicial review of the council's grant of permission. Sullivan J, on 7 November 2003, allowed the claim, holding that the committee should B not have resolved to grant planning permission without first satisfying themselves, pursuant to article 8(b) of the Landfill Directive, that the proposal was "in line with" *Waste Strategy 2000*, in particular with the BPEO methodology incorporated in it.

[18] Before Sullivan J, counsel for Mr Blewett, Mr David Wolfe, put this part of his claim in the following way. The council had failed to C comply with their obligations under *Waste Strategy 2000* to carry out an assessment in order to determine whether the proposed landfill met the requirements of the BPEO in producing the best practicable environmental option for the waste stream(s) in question. In finding in favour of Mr Blewett on this issue, the judge held that: (1) the policies as to the BPEO in *Waste Strategy 2000* should be read more D prescriptively than in earlier national policy guidance; (2) as to landfill, the Landfill Directive, which he clearly regarded as having been implemented into our domestic law, is more prescriptive than the Waste Framework Directive; (3) planning authorities are required, when deciding on permission, not to grant it unless the proposal is "in line with" the policies relating to the BPEO incorporated in *Waste Strategy 2000*; (4) in any event, E the council's consideration of the BPEO had been "seriously flawed"; and (5) even if it had not been, he would not exercise his discretion to refuse relief.

[19] Sullivan J granted permission to appeal "given the importance of the issue: what weight must be given to *Waste Strategy 2000* in the light of the Landfill Directive?". Accordingly, the main issue in the F appeal, as I have said, is whether Sullivan J was correct to hold that compliance with the BPEO policies and methodology is a condition precedent of the grant of waste planning permission in a landfill case. This may be divided into two sub-issues: (1) whether and, if so, how has the relevant provision of the article 8(b) of the Landfill Directive become part of our domestic law; and (2) if it has, whether it imposes a more rigorous obligation with reference to the BPEO in relation to landfill decisions on G waste planning and/or control authorities than the relevant provision of the article 7 of the Waste Framework Directive had previously imposed. In addition, as I have said, there are two further subsidiary issues: (3) whether, in any event, the council's consideration of the BPEO in the case had been so flawed as to be unlawful; and (4) even if it had, whether Sullivan J should have refused relief, or this court should refuse relief, in H the exercise of discretion.

Law and policy

[20] Before considering the competing submissions on those issues, I should outline the relevant law and the policy documents that, in the circumstances, have given rise to them.

Waste Framework Directive

[21] The starting point is the Waste Framework Directive. Articles 3, 4 and 5 of the directive set out what article 7 of the directive calls "the objectives". These objectives have been made part of our domestic law by regulation 19 of, and Schedule 4 to, the Waste Management Licensing Regulations 1994 (the 1994 Regulations)[1] under powers conferred by section 2(2) of the European Communities Act 1972 and other Acts. Article 3 specifies the measures for encouraging the prevention or reduction of waste, its recovery and its use. Article 4, which articulates the central obligation under the directive, deals with recovery or disposal of waste. It provides, so far as material that:

> Member States shall take the necessary measures to ensure that waste is recovered or disposed of without endangering human health and without using processes or methods which harm the environment, and in particular:
> - without risk to water, air, soil and plants and animals,
> - without causing a nuisance through noise or odours,
> - without adversely affecting the countryside or places of special interest...

And article 5 of the directive provides:

> 1. Member States shall take appropriate measures, in cooperation with other Member States where this is necessary or advisable, to establish an integrated and adequate network of disposal installations, taking account of the best available technology not involving excessive costs. The network must enable the Community as a whole to become self-sufficient in waste disposal and the Member States to move toward that aim individually, taking into account geographical circumstances or the need for specialised installation for certain types of waste.
>
> 2. The network must also enable waste to be disposed of in one of the nearest appropriate installations, by means of the most appropriate methods and technologies in order to ensure a high level of protection for the environment and public health.

[22] Article 7 requires competent authorities to draw up waste-management plans in order to attain those objectives. It provides:

> 1. In order to attain the objectives referred to in Articles 3, 4 and 5, the competent authority or authorities referred to in Article 6 shall be required to draw up as soon as possible one or more waste management plans. Such plans shall relate in particular to:
> - the type, quantity and origin of waste to be recovered or disposed of,
> - general technical requirements,
> - any special arrangements for particular wastes,
> - suitable disposal sites or installations.
>
> Such plans may, for example, cover:
> - the natural or legal persons empowered to carry out the management of waste,
> - the estimated costs of the recovery and disposal operations,

[1] SI 1994/1053

> – appropriate measures to encourage rationalisation of the collection, sorting and treatment of waste.
>
> 2. Member States shall collaborate as appropriate with the other Member States concerned and the Commission to draw up such plans. They shall notify the Commission thereof.
>
> 3. Member States may take the measures necessary to prevent movements of waste, which are not in accordance with their waste management plans. They shall inform the Commission and the Member States of any such measures.

[23] In *Commission of the European Communities* v *Greece* C-387/97 [2000] ECR I-5047, the European Court, in para 75 of its judgment, described a waste-management plan within the meaning of article 7 of the directive as "a comprehensive programme with a view to attaining certain objectives". And, in the recent decision of the court in *Commune de Braine-le-Château and Tillieut* v *Région Wallonne* C-53/02 and C-217/02, unreported 1 April 2004, Advocate General Mishko, in para 47 of his opinion, referred to it as "a policy framework… which need not necessarily describe in minute detail all aspects of current and future waste disposal management, including sites". And, as the court observed, in para 30 of its judgment, they may not necessarily be determinative:

> management plans cannot in all cases be the only factor which determines the exact location of waste disposal sites, inasmuch as the final decision concerning location in some circumstances depends on the relevant rules relating to land-use planning and, in particular, the consultation and decision-making procedures implemented pursuant to Council Directive 85/337/EEC of 27 June 1985 on the assessment of the effects of certain public and private projects on the environment… as amended…

[24] As to waste-planning authorities, paras 3(1) and 1 of Schedule 4 to the 1994 Regulations provide that a waste planning authority are a "competent authority", the "specified functions" of which include "determining an application for planning permission". The effect of this is to tie articles 3, 4 (4(1)(b) in particular) and 5 into the domestic planning process. When a waste-planning authority determine an application for planning permission, they are a competent authority discharging one of their specified functions. By paras 2(1) and 4 of Schedule 4, they are required to discharge such function "insofar as… [it] relate[s] to the recovery or disposal of waste" in accordance with the objective in para 4(1)(b), namely "implementing, so far as material, any plan made under the plan-making provisions" (in this instance, *Waste Strategy 2000* and any regional or local plans).

[25] As to the "competent authority or authorities" – note the plural – to draw up these waste-management plans, article 6 of the Waste Framework Directive requires member states to establish or designate them. In England and Wales, this has been implemented by paras 2 and 3 of Schedule 4 to the 1994 Regulations. Table 5 in para 3 includes in the list of competent authorities "any planning authority" in respect of the taking of any "specified action", which, by para 1, includes any planning authority when considering

applications for planning permission for waste proposals. And article 9 of the directive requires "for the purpose of implementing" articles 4, 5 and 7, establishments or undertakings carrying out waste-disposal or recovery operations to obtain permits therefore from the designated competent authority, thus tying the grant of a permit to the plan, if there is one. A

[26] Article 7 of the Waste Framework Directive, as substituted in 1999[1] in its requirement of competent authorities to draw up waste-management plans in order to attain the objectives referred to in articles 3, 4 and 5 of the directive, was, as I have said, expressly made part of our domestic law by section 44A of the EPA. Section 44A provides, so far as material: B

> (1) The Secretary of State shall... prepare a statement ("the strategy") concerning his policies in relation to the recovery and disposal of waste in England... C
>
> ...
>
> (4) ... the strategy must include –
>
> (a) a statement of the Secretary of State's policies for attaining the objectives specified in Schedule 2A to this Act;...
>
> (b) provisions relating to each of the following... – D
>
> (i) the type, quantity and origin of waste to be recovered or disposed of;
>
> (ii) general technical requirements; and
>
> (iii) any special requirements for particular wastes.

[27] The objectives set out in paras 1 and 2 of Schedule 2A are, in substance, the objectives set out in parts of articles 3, 4 and 5 of the directive, and referred to as such in article 7. E

[28] It should be noted that section 44A(4)(a) of the EPA is derived from article 7 and that section 44A(4)(b) deals only with three general objectives, not with installations and suitable disposal sites, which are dealt with in Schedule 4 to the 1994 Regulations. Regulation 19 of those regulations provides that "Schedule 4 (which implements certain provisions of the... [Waste Framework Directive] shall have effect". F Paragraph 2(1) of Schedule 4 to the 1994 Regulations, which clearly applies to both strategic plan-making and planning control, requires "the competent authorities", for this purpose, waste-planning and control authorities to:

> discharge their specified functions, insofar as they relate to the recovery or disposal of waste, [sic] with the relevant objectives. G

It looks as though the draftsman may have mistakenly omitted some words after "waste," possibly "in accordance with", or "in compliance with", or more likely, "in line with", the critical words to be used in article 8(b) of the Landfill Directive: see [38] below. That the last were the missing words is suggested by the Department of the Environment in H para 1.47 of its Circular 11/94, issued on 19 April 1994, drawing attention to, and summarising, these provisions of the 1994 Regulations.

[1] See Council Directive 91/1/156/EEC

A The general duty in paragraph 2(1) means that in exercising their specified functions, authorities must always consider the objectives of the Directive and aim to determine decisions and permit conditions *in line with* them...

(Emphasis added.)

It added, in para 1.48 – as this court was to hold in *R (on the application of* B *Thornby Farms Ltd)* v *Daventry District Council* [2002] EWCA Civ 31; [2003] QB 503[1], and consistently with the formulation of the European Court in *Belgium* v *Traen* C-372/85 [1987] ECR 2141, and *Comitato di Coordinamento per la Difesa della Cava* v *Regione Lombardia* C-236/92 [1994] Env LR 281, in paras 12 and 13 – that the duty was objective, but not absolute. It was a duty "to have due regard to the objectives" mentioned in recital 3 to, and article 4 of, the directive.

C The duty imposed by paragraph 2(1) of Schedule 4 does not remove authorities' discretion in exercising their functions. The objectives are not framed in sufficiently precise terms for that. Account should also be taken of the decision of the European Court in the *Lombardia* case... which has determined that Article 4 should be considered as an objective rather than an absolute requirement.

D [29] Paragraph 4(1), (2) and (3) of Schedule 4 to the 1994 Regulations expressly repeats the objectives set out in articles 3, 4 and 5 of the directive, calling them "relevant objectives" for the purpose of the schedule, and adding, in para 4(1)(b), "implementing, so far as material, any plan made under the plan-making provisions", as defined in section 44A of the EPA, E which, as I have said, is derived from article 7 of the Waste Framework Directive in its provision for the preparation by member states of waste-management plans in order to attain those objectives.

[30] The effect of all this would appear to be that the inclusion, by para 4(1)(b), of plans under the plan-making provisions as a relevant objective under para 4(1) in relation to the disposal or recovery of F waste, although not identified as such in the directive, results in article 7 becoming an obligation upon the state to draw up waste-management plans "to attain" the objectives in articles 3, 4 and 5. That this is so, had the support of this court in *Thornby*, in which, in [9], Pill LJ held, with the agreement of Robert Walker LJ, as he then was, and Laddie J, that there is no material difference between the wording of article 4 of the directive G and the wording of para 4(1) of schedule 4 to the 1994 Regulations. This includes the provision, in para 4(1)(b), for "implementing, so far as material, any plan made under the plan-making provisions", which implementation includes, as provided by para 1 of Schedule 4, planning applications under Part II of the Town and Country Planning Act (TCPA) and section 44A of the EPA. This provision, as Mr James Maurici, on behalf H of the Secretary of State, observed, also reflects the wording of article 9: "[f]or the purposes of implementing Articles 4, 5 and 7, making such implementation an objective for relevant authorities in carrying out their specified functions."

[1] [2002] 2 PLR 21

[31] As will appear, the importance of the nature or strength of this implemented obligation has relevance to the first two issues in this appeal, namely whether article 8(b) of the Landfill Directive has been implemented into, or is otherwise part of, our law, and, if it has or is, whether it imposes any stronger obligation than that resulting from the implementation of the Waste Framework Directive.

[32] It would appear that, to the extent where, as here, the UK has not prepared local waste-management plans in accordance with article 7 of the Waste Framework Directive, it is in breach of its Treaty obligations: see para 22 of *Commission of the European Communities* v *United Kingdom* C-35/00 [2002] ECR I-953, read with *Commission of the European Communities* v *France* C-292/99 [2002] ECR I-4097. In *Commission* v *United Kingdom,* the European Court recorded, in para 22:

> The United Kingdom does not dispute its alleged failure to meet its obligations. It admits that, during the period under consideration, it failed to adopt and/or notify waste management plans capable of covering the entire territory of the United Kingdom... It accepts that the Commission has valid grounds for a declaration of failure to fulfil treaty obligations in the application. It states that competent authorities are engaged in rectifying that situation, by replacing local plans with national strategies, which should constitute the appropriate means of satisfying the requirements of... [the *Waste Framework Directive*]...

[33] In *Commission* v *France,* the court ruled, in para 44:

> considering the objectives pursued by the obligation laid down in Article 7(1) of... [the *Waste Framework Directive*], it is clear from the very wording of that provision that the obligation is necessary in order for the objectives set out in Articles 3, 4 and 5 of that directive to be fully attained (see, by analogy, Case C-387/97 *Commission* v *Greece* [2000] ECR O-5047, paragraph 95). Chief among those objectives is the protection of public health and the environment, which is the essence of Community legislation relating to waste. That is the reason why, according to the case-law, a failure to fulfil the obligation to draw up waste management plans must be regarded as serious, *even if the failure relates to only a very small part of a Member State's territory, such as a single department... or a single area within a valley...*

(Emphasis added.)

Landfill Directive

[34] The Landfill Directive of 1999 was introduced, as Sullivan J correctly identified in [98] of his judgment, to deal with a particular method of waste disposal – landfill – which, as I have said, is at the bottom of the waste hierarchy, that is to say, all other things being equal, it is the least preferred option. One of its purposes was to discourage the unnecessary use of landfill as a method of waste disposal.

[35] The need for such discouragement was well described by Dr Watson, in paras 6 to 8 of her witness statement, on behalf of Friends of the Earth:

> 6. Around 400 million tonnes of waste are produced in England and Wales each year. Of this waste 106 million tonnes is waste produced by

> A industry, commerce and households. Most of the waste produced in England and Wales goes to landfill. .... According to DEFRA about 75 per cent of household waste is landfilled each year...
>
> 7. Landfill, as a means of waste disposal is at the bottom of the waste hierarchy due to the impacts that landfills have on both the environment and society. FOE's concerns about landfill include particularly the following:
>
> B
> - Land filling materials is a waste of resources – as materials that could be re-used or recycled are buried instead.
> - Land filling materials also exacerbate climate change, because when materials are buried more fossil fuels are used to replace the materials, through mining, manufacturing and transportation around the globe.
>
> C
> - Landfill sites are also a major source of methane gas – which is a very potent greenhouse gas.
> - Landfill sites can cause serious pollution of rivers, streams and groundwaters.
> - Landfill sites can also create a nuisance to local communities through the generation of traffic, noise, dust, pests, litter and odours.
>
> 8. As a result of these problems the... Landfill Directive was agreed with D the aim of reducing the degree to which Member States rely on landfill as the principal means of waste disposal.

[36] Thus, recital 3 to the Landfill Directive included among its purposes encouragement of the prevention, recycling and recovery of waste and obviating the wasteful use of land, and recital 2 indicated that it was directed to ensure that, in future, only safe and controlled landfill E operations should be carried out. As the Friends of the Earth have pointed out in its witness statement, the Secretary of State has, in his guidance on this directive, described it as a "step change in the way we dispose of waste in this country...".

[37] As I have indicated, this appeal is concerned with article 8, and more particularly, with article 8(b). The article as a whole is headed "Conditions F of the Permit". However, in considering the effect, in the context of a planning application for a landfill project, of the article 8(b) requirement for the project to be "in line with" the relevant waste-management plan, it is important to distinguish it from article 8(a), which is concerned with the issue of landfill permits.

[38] The two paragraphs are introduced in the same way, and provide, G so far as material:

> Member States shall take measures in order that:
>
> (a) the competent authority [namely the Environment Agency[1] does not issue a landfill permit unless it is satisfied that:
>
> (i) without prejudice to Article 3(4) and (5), the landfill project complies with all the relevant requirements of this Directive...
>
> H (b) the landfill project is *in line with* the relevant waste management plan or plans referred to in Article 7 of... [the Waste Framework Directive]. ...

(Emphasis added.)

[1] See Landfill (England and Wales) Regulations 2002

[39] As Mr Christopher Katkowski QC, on behalf of Derbyshire Waste, A
and Mr David Wolfe, on behalf of Mr Blewett, observed, it is important to distinguish between these two provisions. Article 8(a), which has been expressly made part of our law[1] obliges the Environment Agency to take measures not to issue a landfill permit unless it is satisfied of various conditions, including compliance with all the relevant requirements
of the directive. Article 8(b), which, it is common ground, has not B
explicitly been made part of our law, requires a waste-planning or control authority to take measures in order that a landfill project is "in line with" the relevant article 7 waste-management plan or plans. Thus, a landfill permit under article 8(a) is not the same thing as a planning permission, with which article 8(b) is in part concerned: they are
two different consents granted by different authorities. It is, therefore, C
important not to conflate the provisions of article 8(a) and (b), as Sullivan J may have done in [92] of his judgment, by reading them together so as to impose an obligation on waste-planning authorities not to grant a planning permission unless they are satisfied that the proposal is "in line with" *Waste Strategy 2000.* As can be seen, article 8(b) does not
say that. D

[40] Before leaving the Landfill Directive, I should mention recitals 9 and 18, and articles 1, 3(3) and 3(4), applying the principles of the Waste Framework Directive where the Landfill Directive is excluded, and the introduction to article 9, supplementing the provisions in article 9 of the Waste Framework Directive for the grant of permits for implementing any
of the objectives in articles 4, 5 and 7. Recitals 9 and 18 provide: E

> (9) Whereas Member States should be able to apply the principles of proximity and self-sufficiency for the elimination of their waste at Community and national level, in accordance with... [the Waste Framework Directive], whereas the objectives of this Directive must be *pursued and clarified* through establishment of an adequate, integrated network of disposal plants based on a high level of environmental protection;
>
> ... F
>
> (18) Whereas because of the particular features of the landfill method of waste disposal, it is necessary to introduce a specific permit procedure for all classes of landfill in accordance with the general licensing requirements already set down in... [the Waste Framework Directive]...

(Emphasis added.) G

[41] Article 1 of the Landfill Directive, which is headed "Overall objective", could not make more clear its supplementary role to the Waste Framework Directive. It begins:

> With a view to meeting the requirements of... [the Waste Framework Directive], and in particular Articles 3 and 4 thereof, the aim of this
> Directive is, by way of stringent operational and technical requirements H
> on the waste and landfills, to provide for measures, procedures and guidance to prevent or reduce as far as possible negative effects on the

[1] Pollution and Control (England and Wales) Regulations, regulation 10(2), when read with the Landfill (England and Wales) Regulations 2002

> environment... from land filling of waste, during the whole life-cycle of the landfill.

Article 3.3, under the heading "Scope", provides:

> Without prejudice to... [the Waste Framework Directive] Member States may declare at their own option, that the deposit of non-hazardous waste... can be exempted from... [certain] provisions in... this Directive.

And article 3.4 provides, under the same heading:

> Without prejudice to... [the Waste Framework Directive] Member States may declare at their own option... [certain specified provisions] of this Directive not applicable to: [certain types of landfill sites]...

BPEO and Waste Strategy 2000

[42] I should deal next with the BPEO, which has featured in our planning, including waste-planning, processes long before its present articulation and application in *Waste Strategy 2000*. It had its origin in a decision of OECD countries in 1976, and was adopted and defined in that year by the 12th report of the Royal Commission on Environmental Pollution. It was subsequently incorporated into the Waste Framework Directive by amendment in 1991, which in turn imposed an obligation on member states to prepare a national waste plan incorporating the BPEO. That obligation was anticipated in England and Wales by PPG 10: Planning and Waste Management, issued in 1999, and given effect to by *Waste Strategy 2000*.

[43] As I have said, the draftsmen of *Waste Strategy 2000*, in para 3.4 and annex D, the "Glossary of Terms", have adopted the Royal Commission on Environmental Pollution's definition, namely:

> a BPEO is the outcome of a systematic and consultative decision-making procedure which emphasises the protection and conservation of the environment across land, air and water. The BPEO procedure establishes, for a given set of objectives, the option that provides the most benefits or the least damage to the environment as a whole, at acceptable costs, in the long term as well as in the short term. ...

[44] As to the level of decision making to which the BPEO applies, the many references to it in *Waste Strategy 2000* appear to give mixed messages, but this may be more a product of the different contexts in which it is presented than of general ambiguity. The general notion, set out in Chapter 4 of Part 1, establishes broad principles for action by all sections of society from the government down. However, many references to the BPEO in that part of the document, of which the following, in Chapter 4 of Part 1, are examples, appear to apply more to strategic, national, regional and local applications of the principle, and the relationship between them, rather than to give guidance for decisions on planning applications.
At p40, under the head "Delivering Change":

> Decisions on waste management, including decisions on suitable sites and installations for treatment and disposal, should be based on local assessment of the Best Practicable Environmental Option.

At p41, in para 4.4:

> The right way to treat particular waste streams cannot be determined simply. The objective is to choose the Best Practicable Environmental Option (BPEO) in each case. BPEO varies from product to product, from area to area and from time to time. It requires waste managers to take decisions which minimise damage to the environment as a whole, at acceptable costs, in both the long and the short term. …

At p42, in para 4.5:

> In determining BPEO we will expect those making decisions to take account of three key considerations:
> - the waste hierarchy. Within the hierarchy, the Government and the National Assembly do not expect incineration with energy recovery to be considered before opportunities for recycling and composting have been explored
> - the proximity principle requires waste to be disposed of as close to the place of production as possible. This avoids passing the environmental costs of waste management to communities which are not responsible for its generation, and reduces the environmental costs of transporting waste
> - self-sufficiency. The Government believes that waste should not be exported from the UK for disposal. Waste Planning Authorities and the waste management industry should aim, wherever practicable, for regional self-sufficiency in managing waste
>
> …
>
> 4.7 The Government and the National Assembly now look to key contributors to take action as set out in the following sections.

At p44, in para 4.13, setting out the responsibilities of waste planning authorities:

> Waste Planning Authorities are responsible for identifying suitable sites for waste treatment or disposal installations. The Government and the National Assembly look to the Waste Planning Authorities to:
> - take full account of the policies described in this strategy, in particular:
> - the importance of establishing the BPEO
> - the importance of taking an integrated approach to waste management
> - the need to move substantially away from landfill towards recycling, composting and energy from waste
> - in England, to ensure consistency with the quantity of tradable landfill permits available and with statutory performance standards for recycling
> - implement planning policy guidance fully and quickly – PPGs 10 and 11 in England and PPG (Wales) in Wales – together with any relevant Regional Planning Guidance
> - make realistic assessments of likely future requirements for the number, type and siting of waste treatment facilities in their area in the light of this waste strategy, proposals for development (eg new housing and commercial centres)
> - promote informed debate with the public and businesses in their area about the need for waste management facilities and the options available to produce the Best Practicable Environmental Option
> - work with the Environment Agency to ensure that planning and licence conditions are complementary and effective and to ensure timely decisions, twin tracking where possible

[45] However, Chapter 3 of Part 2 of the *Strategy*, under the heading "The decision-making framework", in its focus on who is responsible for applying the technique and the manner of its application, gives another picture. It has two main themes. First, it sets out its policy of the BPEO as its recommended technique for the making of waste-management decisions. Second, it applies that policy to the waste-management process looked as a whole. Thus, it identifies the BPEO procedure as "a framework" that "local authorities must follow… [w]hen taking waste management decisions on suitable treatment options, sites and installations", and that other decision makers, such as business waste managers, should regard as "a guide". In para 3.1, it emphasises the strategic nature of the exercise, but considered at all levels: national, regional and local:

> The Government and the National Assembly for Wales believe that the most effective waste management decisions can be taken by adopting an integrated approach to waste management. Integrated waste management can be considered to be a number of key elements working in concert, in particular:
>
> - *recognising each step in the waste management process as part of a whole* – decisions should take account of the collection, transport, sorting, processing and recovery or disposal of wastes;…
> - *involvement of all key players* – an integrated approach to waste management should also define the contributions which all interested parties (which might include waste producers and managers, waste re-processors, waste regulators, waste management planners, community groups, consumers and householders, and Government) can make in the development and achievement of common goals and objectives
> - *a mixture of waste management options* – those planning the management of significant quantities of various wastes should avoid over-reliance on a single waste management option. It is unlikely that one approach will represent the Best Practicable Environmental Option (BPEO) for all elements of the waste stream
> - *formal and informal partnerships* – especially between those organisations obligated with legal responsibilities for managing waste that they generate or that arises in their areas. In particular this means Waste Collection and Waste Disposal Authorities within a particular area. Local authorities within a region who will also need to take a collective view of the more strategic regional implications of their various policies.

[46] Paragraph 3.2, in common with other passages in the document, stresses the importance of the BPEO methodology at local level in terms capable of encompassing, not only the development of local waste plans but also of their application as a matter of planning control. It states:

> The following section gives some further detail on determining BPEO. Subsequent sections in this Chapter consider the impact of land use planning, waste management licensing, integrated pollution prevention and control regulation, the Duty of Care, which all have a role to play in determining and implementing optimum waste management solutions across England and Wales. Decisions on how to treat or dispose of waste should be taken locally, taking account of local circumstances and nature of particular waste streams. When taking waste management decisions on suitable treatment options, sites and installations, local authorities must follow the framework set out

> below. This framework should also act as a guide for other decision-makers, A
> including business waste managers.

[47] Paragraph 3.9 attempts to give some guidance as to weighting of the various options in the BPEO waste hierarchy, when determining the mix of options to provide BPEO. Although ostensibly, and no doubt primarily, a strategic exercise, some of its notions in simple form are well-known tools
in the balancing exercise so often necessary for decision making when B
considering an application for planning permission, whether in a waste context or otherwise:

> The judgement about which mix of waste management options provides the BPEO can be resolved by analysing the trade-offs between objectives
> or criteria. This can show the extent to which one objective is sacrificed in C
> order to achieve another (for example, how much costs could rise to reduce the impact on global warming). Formally, this can be resolved using decision techniques such *multi-criteria analysis* (MCA). These entail the systematic modelling of decision-makers' preferences, to resolve the choice between several options involving a number of objectives or criteria. By aggregating disparate information onto a common index of value they provide a rational
> basis for classifying the choices. D

[48] With that over-long introduction to the BPEO, I should now set out in a little more detail its main principles, which, among other things, rate landfill as the least favoured manner of treatment of waste. I draw from paras 3.3 and 3.5 – 3.10 of Chapter 3 of Part 2 of *Waste Strategy 2000*, the
three main principles of the BPEO: namely "the precautionary principle", E
"the proximity principle" and "the waste hierarchy":

> **The precautionary principle**
> where there are threats of serious or irreversible damage, lack of full scientific certainty shall not be used as a reason for postponing cost-effective measures to prevent environmental degradation
> **The proximity principle** F
> waste should generally be disposed of as near to its place of origin as possible
> ...
> **The waste hierarchy**
> a conceptual framework, which acts as a guide...
> - the most effective environmental solution is often to reduce the generation of waste – *reduction* G
> - products and materials can sometimes be used again for the same or a different purpose – *re-use*
> - value can often be recovered from waste, through *recycling, composting or energy recovered*
> - only if none of the above offer an appropriate solution should waste be disposed of.

H

[49] In drawing on passages from *Waste Strategy 2000* as one source of articulation of the BPEO, I have anticipated some of the *Strategy's* content as "the relevant waste management plan" for the purposes of article 7 of the Waste Framework Directive. Paragraph 5 of Part 1 of the *Strategy*, which is headed "Status of the strategy", states:

A This strategy is a waste management plan for England and Wales under the EC Waste Framework Directive... implemented by Section 44A of the Environmental Protection Act 1990...

The effect, as I have indicated in [30] of this judgment, is that para 4(1)(b) of Schedule 4 to the 1994 Regulations should be read in this context as if it read "implementing, so far as material, the Waste Framework Directive in B the form of *Waste Strategy 2000*".

[50] *Waste Strategy 2000* provides, in Chapter 4 of Part 1, that each waste-planning authority, which are usually the local planning authority responsible for the making and enforcement of local planning, including waste management, policy, should prepare a local waste-management C plan taking into account, *inter alia*, the BPEO and planning policy guidance on waste in the form of PPGs[1]. As I have said, at the material time, the council, in their capacity as a waste-planning authority for Derbyshire, had not done that.

[51] The general relationship of *Waste Strategy 2000* and the two directives from which it derives its force within our planning process D is to be found via para 4 of Schedule 4 to the 1994 Regulations, implementing, *inter alia*, the obligation on local waste-management authorities, imposed by article 7 of the Waste Framework Directive, to draw up a waste-management plan. Such authorities, when considering a planning application for use of a site for waste-management purposes, must approach their decision as required by sections 54A and 70(2) of the E TCPA, namely with regard to, and in accordance with, the development plan unless material considerations indicate otherwise. That is because the 1994 Regulations, read together with regulation 10(4)(b) of the Pollution Prevention and Control (England and Wales) Regulations 2000, confine the grant of a permit for the carrying out of a specified waste-management activity at an installation to a site in respect of which, where necessary, F there is planning permission for it. Any local waste-management plan drawn up, or which should have been drawn up, incorporating the BPEO policies pursuant to article 7, is at the very least a material consideration for this purpose. But the question is whether, notwithstanding what would otherwise be a proper exercise of planning judgment, the BPEO considerations might trump the development plan and other material G considerations, as Mr Wolfe has contended. Sections 54A and 70(2) of the TCPA provide:

54A Where, in making any determination under the planning Acts, regard is to be had to the development plan, the determination shall be made in accordance with the plan unless material considerations indicate otherwise.

H 70(2) In dealing with... [an application for planning permission] the authority shall have regard to the provisions of the development plan, so far as material to the application, and to any other material considerations.

[1] PPGs 10 and 11 in England and Wales and PPG Wales in Wales

Issues on appeal A

[52] As Mr Wolfe helpfully summarised it, the central thrust of Sullivan J's judgment was that planning permission should be granted for a landfill site only if it is the BPEO, as incorporated in *Waste Strategy 2000,* for dealing with the waste stream involved. The main issue in the appeal is whether he was correct to regard the BPEO exercise and outcome
as a precondition of planning permission in that way, or whether, as B
Mr Katkowski and Mr Maurici, who appeared for the Secretary of State, put it, the answer in any individual case must turn on the relative weight to be given to the BPEO objectives when considered alongside and against all other material planning considerations. As I have indicated, this issue consists of two overlapping sub-issues, namely: (1) whether and, if so,
how has article 8(b) of the Landfill Directive been made part of our law; C
and (2) if it is part of our law, whether it imposes, in relation to planning applications for landfill sites, a more rigorous obligation under the BPEO, as incorporated in *Waste Strategy 2000,* on waste-planning and/or disposal authorities than article 7 of the Waste Framework Directive in relation to other forms of waste management.

[53] If the judge was wrong on either of those sub-issues, there are, as D
I have said, two further subsidiary issues. They are: whether the council's consideration of the BPEO was so flawed as to be unlawful in any event; and, if so, whether Sullivan J should have refused Mr Blewett relief in the exercise of his discretion.

Issue 1: Implementation of article 8(b) into our domestic law

[54] As I have said, the council did not, in their strategic E
waste-planning role under *Waste Strategy 2000,* prepare a local waste-planning strategy, still less carry out a strategic BPEO assessment for the waste stream(s) in question. In his judgment, Sullivan J's main concern was whether the Landfill Directive objectives imposed wider or more rigorous obligations upon waste-planning authorities in the grant of
planning permissions for landfill sites than did the more general objectives F
in the Waste Framework Directive, and he held that it did. I shall return to the reasons that he gave for that. However, on the threshold point as to whether the relevant provision – article 8(b) – of the Landfill Directive is part of our domestic law, he proceeded, with little or no analysis, on the assumption that it had been implemented. As Mr Katkowski
acknowledged, the judge can perhaps be forgiven for not considering G
more closely the point of implementation since it was not fully argued before him.

[55] Sullivan J's acceptance that article 8(b) is part of our law led him, as a matter of interpretation, in part by contrasting the words "in line with" in article 8(b) with the provisions of the Waste Framework Directive, as
implemented in paras 2(1) and 4(1)(b) of the 1994 Regulations, to conclude H
that it was not sufficient for the council, in deciding to grant Derbyshire Waste planning permission, to keep the requirements of the BPEO, as identified in *Waste Strategy 2000,* "in mind"; they had to be satisfied that the proposal would satisfy those requirements. This is how he reasoned the matter in [98] and [99] of his judgment:

> 98. The 1999 Landfill Directive is concerned with a particular method of waste disposal, landfill, which is at the bottom of the waste hierarchy (that is to say, all other things being equal, it is the least preferred option). The purpose of the Landfill Directive included encouraging the prevention, recycling and recovery of waste and obviating the wasteful use of land (Recital 3), and ensuring that, in future, only safe and controlled landfill operations should be carried out (Recital (2)). In short, it sought to discourage the unnecessary use of landfill as a method of waste disposal.
>
> 99. To this end, Article 8 of the Landfill Directive is more prescriptive than the Framework Directive as implemented by paragraphs 2 and 4(1)(b) of the 1994 Regulations. In ordinary language an obligation to be satisfied that a proposed development is "in line with" a waste management plan, is more stringent than an obligation to keep the objective of implementing the plan, so far as material, in mind. The difference in wording between the two directives, requiring greater weight to be placed upon the waste management plan, is deliberate, having regard to the purposes of the later directive. The words "in line with" admit of some flexibility. They are perhaps less prescriptive than "in accordance with". Moreover, given the complexity of the subject matter and the many factors that may have to be taken into account when taking individual waste disposal decisions, the waste management plan itself may well allow for further degree of flexibility. …

[56] Neither Mr Katkowski nor Mr Maurici suggested that the judge had been wrong to regard article 8(b) of the Landfill Directive as part of our law, and they agreed as to the route by which it had occurred, namely via the implementation by the 1994 Regulations of the Waste Framework Directive. However, Mr Katkowski sought to derive from that route a challenge as to the correctness of the judge's conclusion that article 8(b), via *Waste Strategy 2000* and the BPEO, imposed a more rigorous obligation upon waste-planning and development-control authorities in relation to landfill than in relation to other means of waste treatment.

[57] On the threshold point of implementation, the common starting point of Mr Katkowski and Mr Maurici was that there is no specific domestic provision explicitly implementing article 8(b) of the Landfill Directive. Even the Landfill (England and Wales) Regulations 2002, which came into force in mid-2002, effectively implementing the directive in all or most of its provisions, made no specific reference to article 8(b). The reason for such omission, they submitted, is that there was no need for it, since article 8(b) was no more than a restatement of the requirement in article 9 of the Waste Framework Directive on waste-disposal or recovery operators to obtain permits from the designated competent authority, namely, in England and Wales, from the appropriate waste-planning authority. As a result, they maintained that all that article 8(b) did was to impose an obligation on such an authority to ensure that a landfill project is "in line with" the Waste Framework Directive's article 7 plan as implemented in relation to local waste-planning authorities through para 4 of Schedule 4 to the 1994 Regulations, here via *Waste Strategy 2000.*

[58] Mr Katkowski and Mr Maurici, in elaboration of their common submission on implementation said that the only candidates for implementation of article 8(b) in relation to the role of waste-planning

authorities in determining planning applications are paras 1, 2(1), 3(1) and 4(1)(b) of Schedule 4 to the 1994 Regulations, read together with regulation 10(4)(b) of the Pollution Prevention and Control (England and Wales) Regulations 2000, confining the grant of a permit for the carrying out of a specified waste-management activity at an installation to a site in respect of which, where necessary, there is planning permission for it. But, as those provisions, in their present form, pre-date the Landfill Directive, they submitted that their effect must be that the government considered them apt to implement article 8(b) as they stood, that is, that they did not require amendment in order to effect the transposition. Mr Katkowski maintained, in addition, that it is established in Community law that the result envisaged by a directive does not require implementation into domestic law where it no longer needs to be implemented because the matters covered by it are already in force in a member state, citing the opinion of Advocate General Geelhoed, in para 37, in *Marks & Spencer plc* v *Commissioners of Customs & Excise* C-62/00 [2003] QB 866.

[59] Anticipating briefly the debate on issue (2), Mr Katkowski and Mr Maurici maintained that, on their approach, the nature of a waste authority's duty under paras 2(1) and 4 of Schedule 4 to the 1994 Regulations to discharge their duty in accordance with "the relevant objectives", namely those in articles 3, 4,5 and 7 of the Waste Framework Directive, is governed by our domestic law; in particular, the decision of this court in *Thornby,* in which the court considered, but made little distinction in this context between, an "objective" and a "material consideration". Their point was that, although a waste-planning authority must, as Pill LJ, in *Thornby,* interpreted para 4(1)(b) of Schedule 4 to the 1994 Regulations, "[keep] in mind" the "objective" of implementing *Waste Strategy 2000,* they may still grant permission, in the light of all the material considerations, even if the permission does not implement that objective.

[60] Mr Wolfe submitted that Sullivan J's acceptance that article 8(b) of the Landfill Directive is part of our law and his reasoning and decision on this ground were correct and accorded with obvious common sense. In so submitting, he relied heavily upon the recognition in the strategy and policy documents that, in general, landfill, because it is the least environmentally sound way of dealing with waste, is at the bottom of the list of the hierarchy of options for waste management and should therefore be avoided if possible. He added that if, as Sullivan J held, article 8(b) of the Landfill Directive requires something more than keeping *Waste Strategy 2000* in mind, that can be achieved either by reading domestic legislation to give effect to the directive (*per Marleasing SA* v *La Comercial Internacional de Alimentación SA* C-106/89 [1990] ECR I-4135, or by treating it as a case of direct effect. He said that, however it is done, it cannot dilute the obligation to give effect to the requirements of the directive – thus, it does not matter whether it is by implementation under the 1994 Regulations or otherwise. Domestic law must be construed to give effect to the requirements of the directive.

A [61] As to Pill LJ's treatment, in *Thornby*, of "objectives and "material considerations", Mr Wolfe said that, whether it was correct or not, there is nothing illogical in distinguishing between the former, as something in accordance with which a decision must be taken, and the latter, as merely something that has to be taken into account.

B *Conclusion*

[62] In my view, the short answer to the first issue – whether article 8(b) of the Landfill Directive is part of our law – is that it is part of our law, by the medium, and for the reasons given by Mr Katkowski and Mr Maurici, that is, via the pre-existing 1994 Regulations, in their implementation of the obligation in article 7 of the Waste Framework Directive, "to attain the C objectives... in Article[s] 3, 4 and 5" of it to draw up waste-management plans. It is noteworthy, although not determinative, that that was and is the understanding of the Secretary of State, as expressed in an undated note on the transposition of the Landfill Directive, which states as to article 8(b):

> This requirement has already been transposed in the PPC Regulations D through the duty placed on the Environment Agency not to issue a permit to any waste management activity unless it has already obtained planning permission.

[63] However, even if that were not so, given the role of article 8, and of the Landfill Directive as a whole, as a supplement and a complement to the Waste Framework Directive, to deal with landfill as a particular E form of waste management, there would be a strong common-sense imperative to construe the 1994 Regulations, and any waste-management plan made or amended pursuant to the obligation under article 7 that they implement, as part of the same mechanism for attaining the objectives in articles 3, 4 and 5 of it.

[64] Whichever approach is adopted, I can see no logical basis for F *necessarily* tying the views of Pill LJ in *Thornby* as to the relative meanings of "objectives" and "material considerations" in the context of the incompletely expressed and broad obligations expressed respectively in paras 2(1) and 4(1)(b) of the 1994 Regulations (see [28] above) to the terms of any article 7 waste-management plan. Such a plan or plans may be made and/or amended or supplemented in the light of the possibly G more rigorous requirements of the Landfill Directive or other directives. The meaning in article 8(b) of the words "in line with" a plan required by article 7 cannot depend upon earlier domestic judicial interpretation of domestic legislation effectively implementing article 7. And, as Mr Wolfe observed, if compliance of our domestic law with European law is to depend upon purposive construction, our domestic law must be construed H so as to give effect to the requirements of the two directives read together, not the other way round.

[65] It has to be remembered too that the final content or product of the article 7 obligation on member states turns, not only on effective implementation or domestic purposive construction so as to comply with its broad objective and those of articles, 3, 4 and 5 but also on the content

and wording of their waste-management plans from time to time. Such A plans themselves may vary one from another and internally according to context from prescriptive to the aspirational. In such a "hinged" mechanism for giving domestic effect to European law, domestic judicial formulations of the meaning of a loosely expressed directive, or domestic legislation implementing it in one context, should not be given the prescriptive force in another, albeit related, context, that a later sufficiently B specifically worded directive or domestic statutory provision applying it might command.

[66] However, as signposted in [28] above, and as will appear in my consideration of the next issue, as a matter of construction, Pill LJ's approach in *Thornby* may be as appropriate to the explicit test in article 8(b) of "in line with" as the less explicit formulation in article 7 and paras 2(1) C and 4(1)(b) of Schedule 4 to the 1994 Regulations.

Issue 2: Relative rigour of the Waste Framework Directive and the Landfill Directive

[67] This issue, stated shortly, is whether, as Sullivan LJ held, the meaning and/or effect of the words in article 8(b) of the Landfill Directive, "in line with" the relevant waste-management plan or plans is that D planning permission for a landfill proposal should be given only where it satisfies the policies relating to the BPEO as incorporated in *Waste Strategy 2000*, or whether it is sufficient if the authority have, in the *Thornby* sense, kept the BPEO in mind.

[68] Before I continue any further, I should set out how Pill LJ (with whom Robert Walker LJ, as he then was, and Laddie J agreed), E after a review of the authorities, put the matter in [53] of his judgment in *Thornby*:

> An objective in my judgment is something different from a material consideration. I agree with Richards J that it is an end at which to aim, a goal. The general use of the word appears to be a modern one, in the 1950 edition of the *Concise Oxford Dictionary* the meaning now adopted is F given only a military use: "towards which the advance of troops is directed." A material consideration is a factor to be taken into account when making a decision, and the objective to be attained will be such a consideration, but it is more than that. *An objective which is obligatory must always be kept in mind when making a decision even while the decision-maker has regard to other material considerations.* Some decisions involve more progress G towards achieving the objective than others. On occasions, the giving of weight to other considerations will mean that little or no progress is made. I accept that there could be decisions affecting waste disposal in which the weight given to other considerations may produce a result which involves so plain and flagrant a disregard for the objective that there is a breach of obligation. However, provided the objective is kept in mind, decisions in which the decisive consideration has not been the contribution they H make to the achievement of the objective may still be lawful. *I do not in any event favour an attempt to create a hierarchy of material considerations whereby the law would require decision-makers to give different weight to different considerations.*

(Emphasis added.)

[69] Pill LJ further explained his reasoning in [54], and in expressing his conclusion on the facts in [64]:

> 54. That accords with the approach the ECJ had in mind when the Court used the expression "due regard" in *Traen* and when in *Commission* v *Italian Republic* the Court referred to the "margin of discretion in assessing the need for such measures" (paragraph 67) and accepted that the existence of a situation not in conformity with the objective does not necessarily involve a breach of Article 4 (paragraph 68). The above analysis, which substantially agrees with that of Richards J at his paragraph 48, is also consistent with the wording of Article 249 of the Treaty: "A directive shall be binding, as to the result to be achieved,... but shall leave to the national authorities the choice of form and methods."
>
> ...
>
> 64. ... Derbyshire had to decide whether or not to grant planning permission. The director had regard to material considerations in his report. Derbyshire accepted the director's recommendation that the benefits of the proposal outweighed the disbenefits and granted permission. That was a perfectly ordinary and conventional decision unless it breached Article 4 and paragraph 4. As a contribution to achieving the Article 4 objectives it was of little value in circumstances where there was no need for it, but that does not render it unlawful. The objective is not an overriding consideration on these facts. A council with the objectives well in mind may still grant permission for the reasons given in the director's report. This modest extension of landfill is not so contrary to the achievement of the objective that Derbyshire were obliged to refuse permission. The discretion available to them permitted them to take a broader view than merely to ask which of two possible outcomes contributed more to the objective. The obligation is not overriding in that sense.

[70] This is how Sullivan J dealt with the issue in [100] to [106] of his judgment:

> 100. ... Both parts 1 and 2 of... [*Waste Strategy 2000*] must be read as a whole. It is true that it is an important source of guidance which must be taken into account by local planning authorities. But on its face it professes to be more than that: it implements... the requirement within the Waste Framework Directive... as incorporated into law by section 44A of the Environmental Protection Act. ...
>
> 101. Fairly read as a whole, the policies relating to BPEO in... [*Waste Strategy 2000*] are, and are intended to be, more prescriptive than earlier policy guidance. ...
>
> 102. On a fair reading, the *Strategy* does not simply maintain the status quo in policy terms, leaving local planning authorities free to give such weight as they choose to BPEO. One of the main objectives of the *Strategy* is to "deliver change" by placing greater emphasis on the need to choose the BPEO when making waste management decisions.
>
> 103. It is true that Chapter 3 in Part 2 of the *Strategy* applies to waste management decisions by local authorities generally, but contrary to the advice given to members in the Joint Report... it applies with no less force to waste planning authorities when they are taking decisions on planning applications for waste disposal. ... It is for waste planning authorities when deciding whether or not to grant planning

> permission for landfill proposals to ensure that they are "in line" with Parts 1 and 2 of the Strategy.
>
> ...
>
> 106. For these reasons, I conclude that the... [Council's] approach to the status of the policies relating to BPEO in *Waste Strategy 2000* was erroneous in principle because the Joint Report effectively relegated BPEO to a material consideration to be taken into account but to be given such weight as the... [Council] thought fit. Such an approach did not accord with Pill LJ's pre-Landfill Directive and *Waste Strategy 2000 dicta* in... [*Thornby*]. There was no recognition of the... [Council's] duty, post the publication of the *Strategy* and the implementation of the Landfill Directive, not to grant planning permission unless the proposed development was "in line with" the policies relating to BPEO in *Waste Management 2000.*

[71] Derbyshire Waste's case and the Secretary of State's position is that Sullivan J was wrong to quash the planning permission on the ground that article 8(b) introduced a more prescriptive regime for landfill than article 7 had established for waste management generally so that there should be no grant of planning permission for a landfill proposal unless it were the BPEO in the circumstances. Mr Katkowski and Mr Maurici submitted that, in determining such an application, a waste-planning authority may, but are not obliged to, refuse an application for a landfill permit application because they do not consider that it represents the BPEO for dealing with the waste in question. Mr Wolfe seeks to uphold the judge's ruling and his reasoning for quashing the planning permission on this issue. Friends of the Earth takes the same stance. Dr Watson, in para 23 of her witness statement, puts its case in the following way:

> The BPEO methodology (as elaborated in WS 2000, the UK national plan under Article 7 of the Waste Framework Directive) is the key way in which the UK gives effect to its obligations to seek to progress waste decisions up the waste hierarchy by ensuring that no particular waste decision is made unless it can be demonstrated that options higher up the waste hierarchy are not practicable.

Submissions

[72] Mr Katkowski's and Mr Maurici's submissions, in summary, are that: (1) in determining an application for planning permission for a landfill proposal, a waste authority's powers of determination are not so restricted that they can grant permission only if the proposal is consistent (that is, "in line with") the policies set out in *Waste Strategy 2000*, including, in particular, what the *Strategy* has to say about the BPEO; and (2) in any event, *Waste Strategy 2000* does not state that a landfill proposal can be granted planning permission only if it is "in line" with the BPEO.

[73] Mr Katkowski, whose submissions on this issue were largely supported by Mr Maurici, argued that, in relation to the role of waste-planning authorities in determining planning applications, Sullivan J had erred in treating article 8(b) of the Landfill Directive as if its implementation into our law introduced a more prescriptive regime than that of articles 7 and 9 of the Waste Framework Directive. As I have indicated, Mr Katkowski suggested that, since those provisions

and their implementation predated the Landfill Directive, the reality is that when the latter was issued in 1999 the government considered that article 8(b)'s requirement that any landfill project should be "in line with" any relevant waste management plan or plans was no more than a restatement of the obligation to give effect to article 7 plans contained in article 9. Mr Katkowski submitted, therefore, that the words "in line with" in article 8(b) are no more prescriptive than those in article 9 of the Waste Framework Directive tying the grant of permits and planning permissions to the plan.

[74] As to Pill LJ's approach in *Thornby,* Mr Katkowski submitted that it is consistent, in particular in the passages that I have emphasised in [68] above, with that of the subsequent decision of the European Court of Justice in *Braine-le-Château.* By parity of reasoning, he argued that when, as here, a waste-planning authority determine a planning application for the disposal of waste, the obligation on them to keep in mind the objective of implementing the Waste Framework Directive through the medium of *Waste Strategy 2000* does not require them to reject the proposal unless it implements that *Strategy.* Provided that they have kept the objectives in the *Strategy* in mind, there may be some other material consideration, or considerations, to which they conclude they should give decisive weight. In short, Mr Katkowski submitted that the *Strategy,* in its incorporation of the BPEO, is not an overriding consideration in each case.

[75] Mr Maurici, although a party to those arguments, focused more on the issue of the relative "weight" of the BPEO considerations according to whether the proposal was for landfill or for another means of waste management. He submitted that article 8(b) was intended to provide clarity, not to differentiate in waste-management plans between landfill decisions and other types of disposal or recovery decisions. He said that the strategic nature of the waste-planning process prevents an interpretation of article 8(b) as specifying different requirements from the more general ones contained in the Waste Framework Directive. He developed that line of argument by making two further points. First, the obligation in article 8(b) to ensure that a landfill project is "in line with" article 7 plan is an important, but not the only, planning consideration to which local planning authorities must, through para 4 of Schedule 4 to the 1994 Regulations, have regard when making their planning decisions. Second, neither the coming into force of the Landfill Directive nor the issuing of *Waste Strategy 2000* can affect the status to be accorded to the objectives of the Waste Framework Directive, as implemented in the 1994 Regulations, and as interpreted in *Thornby.*

[76] Mr Maurici also identified the practical difficulties in Sullivan J's conclusion – by whatever route he reached it – that article 8(b) provided a more prescriptive test than, in effect, article 9 of the Waste Framework Directive. He pointed out that *Waste Strategy 2000* is, among other things, intended to provide a framework to help local planning authorities to decide on suitable waste-treatment options. They might, for instance, have to assess whether a particular area really needs a landfill facility or an incinerator to meet best its waste-management needs. It would make

little sense, in the context of a decision as to whether to grant planning A permission for one or other option, to attach a different weight to the local waste management plan according to which option they were considering as against other material planning considerations under section 54A of the TCPA. Although, in their consideration of the local waste-management plan itself, the local planning authority would favour other methods of treatment because, under that plan, landfill is the least preferred option, B they should not confuse the issue of the weight to be given to the plan in relation to all the material planning considerations with the relative weight of the different options for which the plan provided. The weight to be attached to the plan in a landfill case is no greater than that for other methods of treatment; it is just that the plan does not, or should not, favour landfill. C

[77] Mr Maurici's submissions led him to characterise the Landfill Directive as a "supplement" to the Waste Framework Directive. He said that it provides, in relation to landfill, more stringent and detailed operational rules for the general rules contained in the Waste Framework Directive, and thus falls within the provision in article 2(2) of that directive for the laying down "by means of individual Directives" of D "[s]pecific rules for particular instances or supplementing those of this Directive on the management of particular categories of waste" (save for certain particularly hazardous forms of waste specified in article 2(1)(b) already covered by other legislation). He referred to the observation of the European Court in para 48 of *AvestaPolarit Chrome Oy* v *Lapin Ympäristökeskus* C-11/07 [2003] ECR I-8725, tying such further directives to E the general regime of the Waste Framework Directive:

> in contrast to what is expressly laid down in respect of the categories of waste listed in Article 2(1)[(1)] of [the] Directive..., the categories of waste which are the subject of individual directives under Article 2(2) remain subject overall to the... Directive, even if individual rules derogating from its provisions may be adopted on certain aspects and supplementary rules may F be adopted with a view to more extensive harmonisation of the management of the waste in question. ...

[78] Thus, Mr Katkowski and Mr Maurici submitted that, provided that a decision maker keeps in mind the "relevant objective" or the "relevant objectives", it may nevertheless make a decision that does not promote it or them, in this case "*Waste Strategy 2000*, because it considers, say, G that it should give some other material consideration or considerations greater weight.

[79] Mr Wolfe disagreed. He suggested that their argument amounted to an impermissible attempt to make European Union law fit our preexisting domestic regime. He submitted that, by article 8(b) of the Landfill Directive, planning permission had to be "in line with" *Waste* H *Strategy 2000* and thus with the BPEO test that it required to be applied. As to the meaning of "in line with", he drew attention to its substitution by the draftsman for the formula in the final council proposal for the Landfill Directive of "is compatible with", suggesting a conscious tightening of meaning.

[80] Accordingly, submitted Mr Wolfe, the first matter for resolution is the proper construction of the requirements of *Waste Strategy 2000* in relation to the BPEO, which, he maintained, Sullivan J correctly held were intended to be more prescriptive than earlier guidance in relation to it, for example, in PPG 10. He said that *Waste Strategy 2000*'s description(s) of itself in parts as "advisory" must be considered, in the context of the document as a whole, and in the circumstance that it was prepared to implement this country's obligations under article 7 of the Waste Framework Directive to make a "plan", which, by article 8(b) of the Landfill Directive, had to be "in line with" or, as he paraphrased it, "in conformity with" the requirements of *Waste Strategy 2000*. As to the weight to be given in the domestic planning process to its requirements, he said that a domestic characterisation of its articulation of the BPEO principle as being merely another objective to keep in mind is insufficient. He agreed, therefore, with Mr Maurici that the status or weight of *Waste Strategy 2000* is not, as suggested by Sullivan J in [99] to [102] of his judgment, dictated by its own terms, but by the Waste Framework Directive, as implemented via para 4 of Schedule 4 to the 1994 Regulations, article 7 of which requires the local plan to deal, as a matter of local planning policy, with "suitable disposal sites or installations".

[81] On the *Thornby* point, Mr Wolfe pointed out that Pill LJ had not been dealing with article 7 plans. He had been concerned only with the matters in article 4 of the Waste Framework Directive to which para 4(1)(a) of Schedule 4 to the 1994 Regulations gives effect, not the provisions of article 7 of the directive, which do not feature in article 4. In the 1994 Regulations, compliance with the article 7 plan is put as an "objective" alongside the matters from article 4. But that is not how the plan sits within the Waste Framework Directive itself. For example, article 9 requires that an establishment requires a permit "for the purposes of implementing articles 4, 5 and 7. Article 7(1) then requires the plan "in order to attain the objectives referred to in Articles 3, 4 and 5...".

[82] Mr Wolfe also pointed out that Pill LJ had not had the benefit of the European Court's decision in *Braine-le-Château*, in which it held, in para 35, that article 7 requires management plans to include matters of some specificity, such as a map specifying the exact location of disposal sites or sufficiently precise location criteria to enable the authority responsible for issuing a permit for landfill project to determine whether "it falls within the management framework provided by the plan". He also referred to para 45 of the judgment in that case, which seemingly (see [37] and [38] above and [87] below) states that article 8(b) of the Landfill Directive forbids the grant of a landfill permit unless the proposal for it is "in line with" the relevant plan.

[83] Accordingly, submitted Mr Wolfe, even if the matters in article 4, considered alone, are simply objectives to be kept in mind (*per Thornby*), the plan (which is separate) is there to "attain the objectives" in question. It thus goes beyond simply keeping them in mind. In the same context, he

said that a member state's margin of discretion on the implementation of a directive goes only to the means of achieving compliance with it, not as to the fact of compliance: see *Traen*, in paras 21 and 22 and *Commission of the European Communities* v *Italy* C-365/97 [1999] I-7773, in paras 4 and 5 and 64-93. A

[84] He suggested that it follows from that analysis of the relationship with local development plans, that where there is conflict between a local development plan and *Waste Strategy 2000*, a proper application of section 54A and the *Strategy* would require a local planning authority to depart from the development plan. B

Conclusions

[85] In my view, the starting point is that, subject to the provisions of articles 7 and 3, 4 and 5 of the Waste Framework Directive and the guidance now given in *Braine-le-Château* as to the provision by article 7 waste-management plans for location of waste-disposal sites, member states are left with a margin of discretion as to their content at various levels. This margin of discretion may extend, not only to the detail of provision but also to how prescriptive it is. Plans will vary, both according to their national, regional or local scale and, in the latter two cases particularly, differing constraints related to the area, existing provision, needs and competing demands for other use of land. It is against that varying backdrop, which militates against undue specificity and prescription, that article 8(b) requires measures to be taken in order that landfill permits and/or planning permission are "in line with" the plans. The European Court, in *Braine-le-Château*, clearly had these considerations well in mind in the following passage, in para 30 of its judgment: C D E

> management plans cannot in all cases be the only factor which determines the exact location of waste disposal sites, inasmuch as the final decision in some circumstances depends on the relevant rules relating to land-use planning and, in particular, the consultation and decision-making procedures implemented pursuant to Council Directive 85/337/EEC of 27 June 1985 on the assessment of the effects of certain public and private projects on the environment..., or Council Directive 96/61/EC of 24 September 1996 concerning integrated pollution prevention and control... F

[86] Second, far from supporting the argument of Mr Wolfe that the objectives in articles 3, 4 and 5, and that in article 7 to prepare plans to attain those objectives, override all other considerations, including material planning considerations, the court's judgment in that passage expressly acknowledges the tying of permissions to the plan(s) through the planning control process. In so doing, it is of a piece with the scheme of the 1994 Regulations that I have described (see [24] above), which, in its implementation of the relevant provisions of the Waste Framework Directive, does so in the context of planning strategy and control, according to whether the issue is one of adequacy of plans or of compliance with them. If the European Court had considered that article 8(b), as properly interpreted, simply trumped member states' domestic planning systems G H

into which, inevitably, it fell to be considered and applied, it would not have so expressed itself. In my view, it removes the basis for Mr Wolfe's argument that to allow for some balancing of planning considerations alongside the requirements of the two directives, would amount to construing the latter to meet our domestic law, rather than the other way round.

[87] Third, as I have said, the meaning of the words "in line with" in article 8(b) cannot depend upon how it was made part of our law. And article 8(b) does not specify the content of the measures to be taken for a landfill permit and/or planning permission necessary to put them "in line with" the plans. The critical question, therefore, is what is meant by the provision's use of those words. *Braine-le-Château* indicates, post-*Thornby*, that article 7 plans should, according to whether their reach is national, regional and local, locate, or provide criteria for the location of, waste-disposal sites. But it provides little, if any, guidance as to whether, in any individual case, a proposal is "in line with" any relevant plan. The European Court, in para 45 of its judgment in that case, referred to the article 8(b) formula as being "in line with", but it did so without elaboration. And, as I have noted, in doing so, it seemingly referred to article 8(b), when it clearly had in mind article 8(a). The difference between them is important: see [37] and [38] above. Article 8(a) imposes a clearly worded prohibition in relation to the grant of a landfill permit, that is, that it should not be issued unless the project "complies with all the relevant requirements of the Directive…": note, not with all the requirements of the plan(s). Article 8(b), which is in broader and more general terms, and applies to both landfill permits and planning permissions, requires steps to be taken in order that the project should be "in line with" the plan(s). It does not, as does article 8(a) in relation to the directive, forbid the grant of a permit or planning permission in the event of non-compliance with the plan(s). This, it seems to me, is another pointer away from article 8(b), having the effect that compliance with the BPEO, incorporated in *Waste Strategy 2000*, is a precondition of the grant of planning permission.

[88] Fourth, the reasoning of Pill LJ in *Thornby* is another pointer away from such a conclusion. I say his "reasoning", as distinct from the notion, contended for by Mr Katkowski, of it legally fixing the content of article 8(b) when it also became part of our law through the medium of the 1994 Regulations. Although Pill LJ was not concerned with article 7, but with article 4, articulating the central obligation in the directive to take necessary waste-management measures to avoid endangering human health and without harm to the environment, the obligations created by the two provisions are inextricably connected. And that is so whether or not a member state has complied with its obligation to prepare the relevant article 7 plan to attain that and other objectives. In the circumstances, the following passage from para 21 of the judgment of the European Court in *Traen*, where, as it happens, there was no article 7 plan, is of relevance to the issue in this case, as it was to Pill LJ's reasoning in *Thornby*: see [41]:

> That provision [ie Article 4] does not… lay down any particular requirement A
> restricting the freedom of the member states regarding the way in which they organise the supervision of the activities referred to therein. Their freedom must, however, be exercised *having due regard* to the objectives mentioned in the third preamble to the Directive and in article 4 thereof, namely the protection of human health and of the environment.

(Emphasis added.) B

[89] I should add that Pill LJ's reasoning in *Thornby* was drawn not only from [4], but also from the possibly incomplete wording of para 2(1) of Schedule 4 to the 1994 Regulations (see [28] above), as the following passage from [52] of his judgment makes plain:

> … Paragraph 2(1) of Schedule 4 to the 1994 Regulations refers to discharging C
> functions with the relevant objective, terminology which readily encompasses all relevant decisions. I do not consider that anything turns in these cases on the different form in which the obligation is stated in the Waste Framework Directive and in the 1994 Regulations.

[90] In my view, Pill LJ's reasoning in [53] and [64] of *Thornby* (see [68]
and [69] above) distinguishing between an obligation to "keep in mind" the D
Waste Framework Directive's objectives and having regard also to material planning considerations is well made, although necessarily somewhat imprecise. It gives the objectives of the directive the status of important considerations, but not necessarily of overriding weight as against all other considerations in a waste planning permission application. In my
view, that is much the same as the notion articulated in article 8(b) that a E
proposal should be "in line with" those objectives, and as clearly intended, if not articulated, in para 2(1) of Schedule 4 to the 1994 Regulations. Richards J, whose judgment in *R* v *Leicestershire County Council, ex parte Blackfordby and Boothorpe Action Group Ltd* [2001] Env LR 2, at p35, Pill LJ cited with approval in *Thornby*, recognised the practical need, even where
strong imperatives for environmentally sensitive waste management are F
involved, to balance them against other and possibly weightier interests when making strategic planning and planning control decisions. This is how Richards J put it, in [49] of his judgment:

> Paragraph 2(1) of Schedule 4 to the 1994 Regulations does not use the language of taking the objectives into consideration. It requires that planning
> decisions be taken "with" the relevant objectives. This may be contrasted G
> with certain other statutory provisions which impose in terms an obligation "to have regard" to specified objectives. … I have come to the conclusion, however, that there is no real distinction between those formulations. What matters is that the objectives should be taken into consideration (or had regard to) *as objectives*, as ends at which to aim. If a local planning authority
> understands their status as objectives and takes them into account as such H
> when reaching its decision, then is seems to me that the authority can properly be said to have reached the decision "with" those objectives. The decision does not cease to have been reached with those objectives merely because a large number of other considerations have also been taken into account in reaching the decision and some of those considerations militate against the achievement of the objectives.

A [91] I agree with that reasoning of Richards J, which forms the backcloth for Pill LJ's treatment of objectives and material planning considerations in this context. The exercise, as Mr Maurici put it in his submissions to the court, is all a matter of weight. It could possibly be expressed as giving objectives "substantial" weight rather than treating them merely as material considerations, as Richards J, in [49] of his judgment, considered, B but did not decide. Or, it could, as I have just suggested, be expressed as treating them as important considerations. One way or another, it can be said that the added focus given by the Landfill Directive and the BPEO policies in *Waste Strategy 2000* to the environmental harm from landfill when compared with other forms of waste management calls for particular attention or weight to be given to the objectives of the Waste Framework C Directive when considering an application for planning permission for a landfill proposal. However, the attainment of those objectives cannot sensibly be the overriding factor, regardless of all the considerations that are material to any individual decision, so as to be a precondition of the operation of a waste management plan and/or planning permission for a waste proposal. However important such objectives are, the tilt D towards their attainment may, as the European Court recognised in *Braine-le-Château*, be reversed by other and more powerful considerations. The machinery provided by sections 54A and 70(2) of the TCPA allow for this contingency in appropriate circumstances and, in doing so, do not, in my view, violate either of the directives.

[92] Accordingly, I would uphold Derbyshire Waste's case on the E second issue, and hold that article 8(b) of the Landfill Directive does not make it a precondition of the grant of planning permission for a landfill proposal that it satisfies the policies of the BPEO, as incorporated in a waste-management plan made pursuant to article 7 of the Waste Framework Directive, such as *Waste Strategy 2000*.

Issue 3: Whether the council's consideration of the BPEO was flawed in F *any event*

[93] There remains the question as to whether the council's consideration of the BPEO in the case was so flawed as to be unlawful in any event. On this question, Mr Katkowski sought to uphold the council's decision, and Mr Wolfe argued in favour of Sullivan J's rejection of it. Mr Maurici made no submissions on the issue on behalf of the Secretary of State.

G [94] As I have said, the council, in their role as waste-planning authority for Derbyshire, had not prepared a waste management strategy in conformity with the Waste Framework Directive, as implemented by the 1994 Regulations, nor any BPEO document that might have been part of such document if it existed. Nevertheless, as I have also said, the relevant obligations in that directive and the Landfill Directive are not H only applicable to the preparation by national and local waste authorities of strategic waste-management plans but also to the decisions made by local planning authorities, usually the same bodies, in their consideration of planning applications for waste projects on particular sites, including installations and landfill tips. Sullivan J correctly recognised, in [105] of his judgment, that the national strategy, *Waste Strategy 2000*, including its

BPEO components, should be taken into account in both plan-making and development control.

[95] However, Sullivan J went on to hold that the council's consideration of the BPEO, as indicated mainly in the council's officers' report and joint report, had been "seriously flawed", and on that account unlawful, regardless of the interpretation to be given to article 8(b) of the Landfill Directive and/or of the weight the law required them to attribute to the policies in *Waste Strategy 2000*. I shall return to the terms in which he so held after summarising and setting out the council committee's consideration of the matter as shown largely in the report and joint report.

[96] There are two overlapping questions as to the effect of the BPEO in the circumstances of this case. The first is whether, notwithstanding the council's failure of their obligation under the Waste Framework Directive to prepare a waste-management plan, they were required to consider, in their planning control capacity in determining Derbyshire Waste's application for planning permission, what such a plan, pursuant to *Waste Strategy 2000*, should have contained, namely the BPEO. Thus, the challenge is as to their planning control decision, not their failure to provide a waste-management plan containing the BPEO provisions. The answer to that challenge may depend upon the obligations imposed by the BPEO. These include their level of generality or otherwise and the extent to which a local planning authority, exercising their planning control functions on a case-by-case basis, could undertake the sort of analysis required for a BPEO exercise, bearing in mind the national, regional and cross-county considerations that may be relevant to the analysis. The second question, assuming that the council were required to consider the BPEO, is whether, in the absence of a local waste-management plan, they did so adequately, bearing in mind that *Waste Strategy 2000* does not say that the BPEO is a precondition of the grant of planning permission.

[97] Whatever the law required of the council in relation to the policies of the BPEO, it clearly regarded them as relevant to their planning decision. For convenience of reference, I repeat the summary that I gave at the start of this judgment of the council committee's reasons, by reference to the report and the joint report for granting permission, namely: (1) the proposal would be unlikely to cause significant harm to the amenities of nearby residents; (2) there would be no shortage of landfill space in the county considered as a whole before 2010 if reduced waste production – about which there was some uncertainty – and landfill targets were achieved; (3) the proposal was not so large as to create a substantial excess capacity; (4) it would, in any event, assist in meeting disposal needs in the south-east of the county where an imminent shortfall of landfill space was expected; (5) the proposals were probably the only satisfactory way of achieving a sympathetic and worthwhile restoration of the site; and (6) the proposal accorded with the relevant development plan policies and national and regional policy and guidance, and had taken into account other material considerations.

[98] As to policy, although, as the report indicated, the council, at that time, had no strategic local waste plan, they had a non-statutory waste management strategy that had been produced in 1999, which, along with its structure plan, reflected national guidance in the form of the BPEO. The structure plan, adopted in January 2001, also acknowledged the strategic principles set out in *Waste Strategy 2000,* and set out three criteria in its waste management policy 1, which were set out and discussed in the report, namely:

> Provision will be made for sufficient sites and facilities to cater for the waste management needs of Derbyshire, having regard to the national, regional and local strategies for waste management. Particular account will be taken of:
>
> 1) the need to pursue objectives which further the aim of achieving sustainable waste management, such as to find the Best Practicable Environmental Option for individual waste streams;
>
> 2) the aim of minimising the overall volume of waste going to landfill disposal and maximising the recovery of value from waste; and
>
> 3) the location and technical suitability of sites.

[99] The director expressed the view, in the report, that the structure plan accorded with national, regional and local waste strategies. He identified and summarised in the report the guidance that he had taken from, *inter alia,* the Landfill Directive and *Waste Strategy 2000,* including the incorporation in the latter of the BPEO and of its three key considerations, namely: the hierarchy principle, the proximity principle and self-sufficiency, and also PPG 10, containing similar policy guidance. In his assessment of the proposal against those and other national policies and guidance and other material considerations, he expressed, in the following terms, the view that it conformed with *Waste Strategy 2000* and with the three key considerations of the BPEO:

> Whilst the application site is particularly accessible from the north-east of the County, the site also has good connection to the M1 Motorway and A38 trunk route to serve the wider needs of Derbyshire and I am mindful of the imminent shortage of landfill space in the south-east of the county. Thus, I consider that land-filling at this site would be in accordance with the key considerations *Proximity Principle* and *Regional Self-Sufficiency* and technically suitable for land-filling as proposed, thereby providing a *Best Practicable Environmental Option* for the disposal of waste with [criterion] 1 of… [Waste Management Policy 1 of the Structure Plan].

He also concluded that it satisfied the other two criteria of need and suitable location respectively and, moreover, expressed the firm view that use of the site for landfill was the only way in which it could be satisfactorily restored.

[100] The director also considered the effect of the *Thornby* decision, correctly stating that the Court of Appeal in that case had not considered that there is an overriding obligation to further the objective in article 4 of the Waste Framework Directive in any individual planning application, but that the local planning authority considering it should "have the objective in mind".

[101] As to the committee's conclusion, based upon the report, that the landfill proposal was the only satisfactory means of restoration of the site, the emphasis given by the report in its general analysis and in its planning conclusions to that important and practical consideration should be noted:

> With respect to the Structure Plan Waste Disposal Policy 2: Waste as a Positive Resource, the proposal provides a means of reclaiming the last remaining area of dereliction at the former colliery. ... I think it unlikely that other methods of reclaiming the site without importing waste, would produce a satisfactory result in landscape terms...
>
> Notwithstanding the availability of alternative sites both currently, and which may become available in the north-east of the County within the Waste Management Plan period..., I consider that there are compelling reasons to accept the infilling/land raising/restoration of the site as submitted to restore the site satisfactorily and conserve and enhance its ecology thereby providing a significant benefit. I consider that there is no realistic likelihood of an appropriate restoration being achieved without the importation of waste in the manner proposed.

Those sentiments were touched on orally in the committee's meeting, as the following part of the minutes records:

> Members, having considered the report and heard the comments made and explanations provided by the officers, generally considered that there were not any substantial planning grounds for refusal of the application. It was felt that the site was in need of improvement but that it would be unlikely to regenerate in a satisfactory manner on its own. An officer explained that satisfactory restoration without use of waste was a technical possibility but was not feasible except at great expense and that no such alternative scheme was likely to be promoted. ...

[102] The joint report returned to the effect of the BPEO and to the impact on its application of the *Thornby* decision in the following passage

> The report to Committee of 11 March [ie the Report] explained the concept of BPEO (ie the option that provides the most benefits at the least damage to the environment as a whole, at acceptable costs, in the long term as well as the short term), and analysed it in the context of this proposal.
>
> The challenge [ie Mr Blewett's judicial review claim] essentially alleges that the Council's treatment of BPEO... was insufficient. In particular, the level of detail that should be taken into account in determining a planning application, including the lack of identification of the specific BPEO for particular waste streams.
>
> The Courts have held that in appropriate cases BPEO is an objective to which planning authorities should have regard as a material consideration. It is for local planning authorities to decide how much weight to attach to it. In this case the waste hierarchy and the proximity principle were considered and reference was made to the relevant Planning Policy guidance, Waste Strategy 2000, Regional Planning Guidance and the Derbyshire Waste Management Strategy. The ES [environmental statement] made reference to the applicant's own waste management strategy and proposed recycling rates. In particular, the report identified the waste hierarchy, the proximity

> principle and self-sufficiency as considerations. It addressed the issues of the targets for reducing, re-using and recovering value from waste and the requirements for landfill capacity for the residual wastes. In the context of Structure Plan policies it identified the use of waste as a positive resource to reclaim this site.

[103] Sullivan J dealt quite shortly with these deliberations, in [107] and [108] of his judgment:

> 107. But the… [Council's] consideration of BPEO was seriously flawed, regardless of the weight that should have been attributed to the policies in the strategy. Mr Evans [counsel for the Council] and Mr Barrett [counsel for Derbyshire Waste] pointed to a number of places in the Report where BPEO was mentioned. I accept that there are frequent references to BPEO in the Report, but merely repeating the acronym, however frequently, and whether or not accompanied by the Royal Commission's definition, is not an adequate consideration of the issues raised by BPEO. If a material consideration is to be taken into account it must first be properly understood. What matters is not the letters BPEO, but the analysis of the issues raised by the concept: the application of the three key elements the waste hierarchy, the proximity principle and self-sufficiency to the particular waste stream(s) which the development is intended to serve. So long as there was both a local (in the North-East Derbyshire Sub-Area) and county-wide shortage of capacity, it was relatively easy to see how the proximity principle might be met. It would appear that this must have been the assumption underlying the environmental statement, since it contained no discussion of BPEO whatsoever. However, once it had been concluded that there was capacity both locally and county-wide up to 2011, the question whether this particular application site would be the BPEO for meeting a shortage of landfill space in the south-east of the county had to be addressed in terms of the three key considerations, including the proximity principle. Beyond referring to the application site's good road connections, and stating that the Director was "mindful of the imminent shortage of landfill space" in the south-east of the county, the report did not address this issue at all. It may well be that this is why the Director did not feel able to conclude that the site was the BPEO in accordance with criterion 1 in the Waste Management Policy in the Structure Plan, merely that it was "a BPEO for the disposal of waste".
>
> 108. I accept that officers' reports should not be read in a legalistic or pedantic manner. If there had been a reasonable attempt to grapple with the issues raised by BPEO in the light of local spare landfill capacity and capacity county-wide for the structure plan period, the use of the indefinite rather than the definite article might have been of little consequence, and reference to it dismissed as mere pedantry. Its use in this Report is, in my judgment, a reflection of the… [Council's] muddled approach to the BPEO issue. Unfortunately, the muddle was compounded, rather than clarified, by the advice given to members in the Joint Report as to the weight that they ought to give the BPEO. Given the importance attached to choosing the BPEO for a particular waste stream in *Waste Strategy 2000*, this was a significant flaw in the decision-making process.

[104] In [110] of his judgment, the judge added that "[i]n deciding whether waste disposal activities are justified a BPEO assessment is… a most material consideration".

Submissions A

[105] Mr Katkowski had three main criticisms of Sullivan J's reasoning. First, he submitted that the absence of a local waste-management plan effectively watered down the effect of the policy statements concerning the BPEO in *Waste Strategy 2000* on determination of individual planning applications to "something in between the aspiration expressed in the *Strategy* and the site and scheme specific details set out B in an individual planning application". He drew attention, in particular to, the provisions of Part 1 of Chapter 4 of the *Strategy*, including, in particular, those that I have set out or summarised in [44] above, suggesting that they were of a piece with the general requirements of article 7 of the Waste Framework Directive, namely concerned with the BPEO at the strategic level, not apt for dealing with C individual applications.

[106] Second, and in the alternative, Mr Katkowski suggested that Sullivan J's approach, in the passages that I have set out from [107] and [108] of his judgment, was itself flawed since it appeared to be based upon the incorrect notion that *Waste Strategy 2000* contains a mandatory requirement for any landfill proposal that it D must be the BPEO in order to obtain planning permission. He submitted that, on the contrary, the *Strategy's* stated policies, including its guidance in relation to the BPEO, are aspirational and fairly broad-brush; they do not descend to the level of detail that would make the grant of permission for a landfill proposal dependent upon it being consistent, or "in line with", the policies set out in the *Strategy;* E and nowhere does the *Strategy* state that a landfill proposal can be granted planning permission only if it is the BPEO in the case in hand. He said that had either been intended, it would have been spelt out in the document.

[107] Third, Mr Katkowski submitted that, in any event, the judge, in his remarks, had not done justice to the council's consideration of the concept of the BPEO and their application to the circumstances of this case, as found in the report and the joint report and the minutes of the F planning committee's meetings. He criticised the judge for treating those reports and minutes as if they were pleadings, referring the court to *Levy* v *Environment Agency* [2002] EWHC 1663 (Admin); [2003] Env LR 11, *per* Silber J, in [119]. He added that, in any event, if the court is with him on the BPEO point, the judge was wrong to find that the council's reasoning had been seriously flawed, in particular in the director of the report's use G of the indefinite article in the passage in the report concluding that the proposal would provide "a Best Practicable Option for the disposal of waste": see [101] above.

[108] Mr Wolfe submitted that the judge's reasoning was the application of a very low threshold and correctly presupposed that a proper BPEO assessment is a necessary precondition of the analysis regardless of its H status and the precise route of its application.

Conclusions

[109] I was initially taken with Mr Katkowski's first submission that, in the absence of a local waste-management plan, a local authority,

A when determining an application for planning permission for a waste proposal, could not be expected to make much of a fist of applying the BPEO policies, and that their failure to do so should not be a ground for impugning their decision. It is undoubtedly the case that, in the context of an individual planning application, an authority would not normally have the same resources for a full and well-informed
B evaluation that they would normally be expected to undertake in their waste-planning capacity. That was very much the approach taken by Mr Thomas Forster, the waste-management consultant of Derbyshire Waste, on whose evidence, in particular, para 4.2 of his witness statement, Mr Katkowski drew in support of his submission on this aspect. On the face of it, further support for that submission can be derived
C from various public documents giving guidance on the application of waste-management policies at the planning application stage. For example, PPG 10, Planning and Waste Management, issued in October 1999, just a year before the publication of *Waste Strategy 2000,* did not suggest that each proposed waste disposal location should have its own individual BPEO assessment, stressing, in paras 3, 17 and 28, the regional derivation
D of the methodology for application at local level, and, in paras 39 to 42, the variety of considerations and wide range of issues involved. And the *Guidance on Policies for Waste Management Planning* prepared for the Department for Transport, Local Government and the Regions in May 2002, makes the point, in paras 4.6, 4.7 and 4.13, that it is difficult to demonstrate application of the BPEO methodology on a
E case-by-case basis.

[110] These and other sources indicate the difficulty for a local authority, when considering a waste planning application, to carry out what, for this purpose, I shall call "a full BPEO" evaluation. However, there is nevertheless a strong case for concluding that local authorities that are confronted, when considering waste planning applications with BPEO
F considerations in the absence of a regional or local waste-management plan, are required, by law, and can reasonably be expected to undertake some BPEO analysis.

[111] First, as to the law, the Waste Framework Directive, in the obligations that it imposes upon "competent authorities" in the exercise of their "specified functions", includes local waste-planning
G authorities determining applications for waste planning permission: see [24] and [25] above. A local waste-planning authority cannot pray in aid their own failure to draw up a waste management plan as an excuse for failing to comply with the obligations cast upon them by the directive. The same applies to the supplementary obligations imposed by the Landfill Directive in relation to waste management
H by way of landfill: see [41] above. True to those implemented obligations, our national waste-management plan, in the form of *Waste Strategy 2000,* clearly calls for the application of BPEO policies, so far as practicable, at the planning control stage, as well as in strategic national, regional and local waste-management planning: see [45] to [47] above.

[112] Second, as to the practicalities, that legal outcome is A
not unreasonable since planning applications in respect of waste are
normally decided by waste-planning authorities, as specifically provided
by paras 1 and 3 of Schedule 4 to the 1994 Regulations. Those authorities
have knowledge of their respective areas' waste policies, whether to
be found in local waste-management plans or in their general
planning development plans. This is a feature that the draftsman of B
PPG 10 (pre-Landfill Directive and *Waste Strategy 2000*) had recognised
in providing, in annex A of para A52, for the BPEO to be considered
on individual planning applications in respect of sites "not previously
identified as having potential for waste management options". From
all of this, it is plain that *Waste Strategy 2000* and the BPEO principles
that it incorporates must, so far as practicable and identifiable in the C
circumstances of any individual case, be capable of application in
the determination of applications for planning permission, whether
or not there is a local waste-management plan to which the planning
authority can have recourse. I would, therefore, reject this argument of
Mr Katkowski.

[113] As to Mr Katkowski's second criticism of Sullivan J's reasoning, D
I do not consider that the judge, when considering this issue in [107]
and [108] of his judgment, had clung to any notion that *Waste Strategy
2000* made it a precondition of planning permission that any landfill
proposal must be the BPEO. To the extent that that may have been an
accurate reflection of his view on issue (2), he expressly discarded it in
his opening sentence in [107]. His two major points were, in [107], the E
inadequacy of the council's treatment, in their officers' reports, of the
BPEO waste hierarchy as applied to the factual circumstances attending
the proposal, and, in [108], their advice that the BPEO was simply a
material consideration, the weight of which was for them to decide. His
"side-swipe", at the end of [107], about the use of the indefinite instead of
the definite article when referring to BPEO was not necessary to either of F
those criticisms, as the first sentence of [108] of his judgment makes clear.
In short, his criticism was that the council, acting on the advice of their
officers, had not demonstrated that they had given sufficient importance
or weight to the BPEO policies on the facts of the case, a criticism that, if
tenable on the facts, cannot be described as seriously flawed as a matter
of law or otherwise. G

[114] Mr Katkowski's third submission, that the judge's criticisms of
the council did not do justice to their consideration and application of the
BPEO policies in the circumstances of the case, is a challenge to the judge's
findings on the facts. In order for it to succeed in this court, it would be
necessary to show that those criticisms, findings of secondary fact, were
so ill-founded as to be perverse in the sense expressed by Lord Fraser in H
G v *G* [1985] 1 WLR 647, at p652, so as to "exceed... the generous
ambit within which a reasonable disagreement is possible". Although
another judge might reasonably have taken a different view after
reading the reports to, and minutes of, the council committee's meetings,
I do not consider that it could be said that this experienced planning

judge's view approaches being perverse in the sense described by Lord Fraser. Accordingly, I can find nothing under this ground of appeal to justify the court in disturbing Sullivan J's ruling that, whatever the true state of the law as to the effect of article 8(b) of the Landfill Directive on the council's obligations under the Waste Framework Directive, the council's reasoning on the application of the BPEO to the circumstances of this application was so seriously flawed as to render their decision unlawful.

Issue 4: Court's discretion

[115] The fourth question is whether the judge, in the exercise of his discretion, should have refused relief in any event.

[116] Mr Katkowski urged upon the court that, if all his submissions thus far did not find favour with the court, it should exercise its discretion not to quash the planning permission. Sullivan J, in [109] and [110] of his judgment, rejected a similar submission made on behalf of Derbyshire Waste:

> 109. The... [Council's] failure to deal adequately with BPO, whether it is regarded as a breach of its obligation to ensure that the grant of planning permission was in line with Waste Strategy 2000, or whether it is viewed more simply as a failure to have regard to a material consideration, does not mean that the planning permission must be quashed. The court has a discretion and I have anxiously considered whether it would be right in all the circumstances to exercise that discretion, given the two-fold justification for the development in the environmental statement to reclaim a despoiled site and to facilitate the disposal of wastes arising in the area. It is clear from the Report and from the Minutes of the Meeting on 11th March 2002 that the Director placed considerable weight upon the first justification: "there was no realistic likelihood of an appropriate restoration being achieved without the importation of waste in the manner proposed". However, it was for members to determine the application. The minutes record that they "generally considered that there were not any substantial planning grounds for refusal of the application. It was felt that that the site was in need of improvement but that it would be unlikely to regenerate in a satisfactory manner on its own".
>
> 110. Given the manner in which BPEO was addressed in the Report and Joint Report it is not surprising that members concluded that there were no substantial planning grounds for refusing planning permission. Since there had been no proper BPEO analysis it is not possible to say whether there would or would not have been a substantial planning objection on this ground, for example because of failure to comply with the proximity principle. Thus it is simply not possible to tell what members' attitudes might have been if there had been a proper analysis of the BPEO issue, including both the weight to be given to, and the content of, the policies relating to BPEO in *Waste Strategy 2000*. In particular, Waste Management Policy 2 in the Structure Plan gives preference to waste disposal proposals that assist in the reclamation of derelict or despoiled land, "where waste disposal activities are justified"... In deciding whether waste disposal activities are justified a BPEO assessment is, for the reasons set out above, a most material consideration.

Submissions A

[117] Mr Katkowski submitted that, even if the reporting officers and the committee had concluded, on what Sullivan J regarded as "a proper analysis", that the proposal was not the BPEO, they could, and would, still have decided, on a proper reading of *Thornby*, that the impossibility of satisfactory restoration of the site without landfill was a material consideration of overriding significance in favour of the grant B of permission.

[118] Mr Wolfe suggested that this argument depended upon a number of false premises, namely: (1) that the court concludes that, regardless of flaws in the council's analysis, they would have granted planning permission anyway, which is not sustainable on the evidence; (2) on the court accepting that *Waste Strategy 2000* and the C BPEO policies that it incorporates are "mere material considerations"; and (3) that the court has a discretion to excuse non-compliance with a directive, which it has not, since article 10 of the EU Treaty puts the court itself under an obligation to take all necessary measures to secure compliance with a directive, thus precluding the exercise of such discretion. D

Conclusion

[119] In my view, this is not an appropriate case for the court to take a different view from Sullivan J as to the exercise of discretion. For the court to do so would amount to a consideration of the planning merits, an exercise that it should not undertake. There is also E force in the various submissions of Mr Wolfe. It is impossible to say whether the council would have taken the same decision had they sensed the greater importance or weight that they should now give to the BPEO hierarchy that puts landfill at the bottom of the list of waste-management options. And the following words of Lord Hoffmann in *Berkeley* v *Secretary of State for the Environment, Transport and the Regions* F [2001] 2 AC 603[1], at p616C-F, with whom the other law lords agreed, militate against the retrospective dispensation of an obligation under a directive in that way:

> A court is... not entitled retrospectively to dispense with the requirement of an EIA [environmental impact assessment introduced to implement a Directive] on the ground that the outcome would have G been the same or that the local planning authority or the Secretary of State had all the information necessary to enable them to reach a proper decision on the environmental issues. ... It is exceptional even in domestic law for a court to exercise its discretion not to quash a decision which has been found to be *ultra vires*: see Glidewell LJ in *Bolton Metropolitan Borough Council* v *Secretary of State for the Environment* (1990) 61 P&CR 343, 353. ... H

[120] Accordingly, I would dismiss the appeal.

---

[1] [2000] 3 PLR 111

A **BUXTON LJ:**

[121] I agree.

LAWS LJ:

[122] I also agree.

Appeal dismissed.

T-Mobile UK Ltd and others *v* First Secretary of State and another

[2004] EWCA Civ 1763

COURT OF APPEAL
Pill, Mummery and Laws LJJ
12 November 2004

Planning permission – Mobile telephone mast – International A
Commission on Non-Ionising Radiation Protection (ICNIRP) guidelines – PPG 8 (Telecommunications) – Development complying with ICNIRP guidelines – Whether PPG 8 policy requiring additional assurances on public health issues where ICNIRP guidelines satisfied

The application of the first respondent (T), for planning permission to B
install and extend certain telecommunications equipment at an existing mast site, was refused by the second appellant local planning authority in June 2003. The purpose of the proposed development was to enable the second and third respondent mobile telephone companies to share the single mast owned by T, placing their own nine antennae and six dishes upon it. These additional antennae and dishes required a more C
substantial mast than the existing one. In their reasons for the refusal, the local planning authority said that the proposed mast and headframes would, owing to their bulk and massing, unreasonably detract from the residential amenity of the nearby dwellinghouses and the amenity of the local facilities, so as to conflict with policies of the local plan. The Secretary of State, by his inspector, dismissed T's appeal against that decision D
upon consideration of matters relating to health risks. The respondent companies' challenge to that decision, under section 288 of the Town and Country Planning Act 1990, was allowed in the court below: the deputy judge concluded that once the inspector was satisfied that International Commission on Non-Ionising Radiation Protection (ICNIRP) guidelines were met, which they were, the inspector did not have to consider health E
matters any further. On appeal, the Secretary of State contended, *inter alia*, that the judge in the court below had misunderstood relevant provisions contained within PPG 8 (Telecommunications), and that the policies that state that if in a given case the ICNIRP health guidelines are met, the planning authority should not have to look further in relation either to actual or perceived health risks, do not preclude the consideration of F
further health aspects and concerns.

Held The appeal was dismissed.

The judge in the court below had been right to conclude that the inspector had misunderstood PPG 8. In deciding that his conclusion, that the proposals provided insufficient reassurance on health, was consistent with government policy, the inspector had been wrong. The proposal complied with ICNIRP guidelines and there was an appropriate certificate

A to that effect. The inspector's conclusion represented a departure from the policy in PPG 8. There was nothing on the facts of this particular case to show why compliance with ICNIRP guidelines was insufficient to allay perceived fears about health issues. There were no exceptional circumstances in this case: see [21].

B **Case referred to in the judgment**

EC Gransden & Co Ltd v *Secretary of State for the Environment* (1985) 54 P&CR 86; [1986] JPL 519

Appeal against a decision of Sir Richard Tucker, sitting as a deputy judge of the Administrative Court

C This was an appeal by the First Secretary of State and Harrogate Borough Council, against a decision of Sir Richard Tucker allowing an application of the respondents, T-Mobile UK Ltd, Hutchinson 3G UK Ltd and Orange Personal Communications Services Ltd, under section 288 of the Town and Country Planning Act 1990, challenging a decision of the first appellant to dismiss an appeal against a decision of the second appellant to refuse D planning permission.

Philip Coppel (instructed by the Treasury Solicitor) appeared for the first appellant, the First Secretary of State.

Christopher Katkowski QC and Galina Ward (instructed by Freshfields Bruckhaus Deringer) represented the respondents, T-Mobile UK Ltd, Hutchinson 3G UK Ltd and Orange Personal Communications Services Ltd.

E *The second appellants, Harrogate Borough Council, did not appear and were not represented.*

The following judgments were delivered.

[1] **PILL LJ:** Laws LJ will give the first judgment.

[2] **LAWS LJ:** This is an appeal, with permission granted by the judge F below, against the decision of Sir Richard Tucker, sitting as a deputy judge of the Administrative Court on 23 June 2004, when he allowed the respondents' application, brought under section 288 of the Town and Country Planning Act 1990, against the decision of an inspector appointed by the appellant, the First Secretary of State.

[3] By his decision letter of 30 December 2003, the inspector had G dismissed the first respondent's appeal against the refusal on 3 June 2003, by Harrogate Borough Council, of its application for planning permission. The application was for a mobile phone development: that is to say, the respondent proposed to install and extend certain telecommunications equipment at Harrogate. There had been an existing telecommunications mast in place on the site since July or August 2001, but that accommodated H the antennae of the first respondent only.

[4] The stated purpose of the development was to enable the second and third respondents to share a single mast with the first respondent, placing their own nine antennae and six dishes upon it. These additional antennae and dishes required a more substantial mast than the existing one.

[5] The reasons given by the local planning authority for refusing permission were cited by the learned judge at the beginning of his judgment:

> The proposed mast and headframes due to their bulk and massing notwithstanding the existing installation would unreasonably detract from the residential amenity of nearby dwelling houses and the amenity of the local facilities such as to conflict with Policies of the Harrogate District Local Plan.

[6] The first respondent's appeal to the Secretary of State's inspector was conducted by the written representations procedure. The first and second respondents both put in full appeal statements, as did the local planning authority. As the judge was to point out, the inspector did not uphold the local planning authority's original reasons for refusing permission; rather, he dismissed the appeal upon consideration of matters relating to health risks. Health matters were raised in the local planning authority statement. They noted, as was the fact, that there was a school building 170m from the site, another school 250m from the site and a third school 420m from the site. The schools had expressed concerns about the proposal on health grounds, and I will return to that shortly.

[7] In order to elucidate the issues arising for our determination, it is convenient at this stage to give a little background concerning the apprehension of health risks in relation to the use of mobile phones and their associated telecommunications equipment. I can do no better than to cite this passage from the judgment of the learned deputy judge:

> 10. Concerns have arisen in recent years about the possible dangers to health arising from the use of mobile phones and the emission of radio frequency from them and from the transmission stations of the kind to which this application refers. Particular concern was expressed about the siting of base stations on or near school premises. Therefore, the Government decided to establish an independent expert group to examine possible effects of mobile phones, base stations and transmissions on health. The group's Chairman was Sir William Stewart. It made a comprehensive enquiry into the problem and published a detailed report in April 2000.
>
> 11. The Report recommended that a precautionary approach to the use of mobile phone technologies should be adopted until much more detailed scientifically robust information on any health risks becomes available. They recommended that the ICNIRP [International Commission on Non-Ionising Radiation Protection] (International) guidelines for public exposure be adopted for use in the United Kingdom, rather than the national guidelines.
>
> 12. The group specifically considered the question of the siting of base stations near schools, and made a number of detailed suggestions, relating to the siting of such stations and to the areas within which the beam of greatest radio frequency intensity should or should not be allowed to fall.
>
> 13. There was a prompt response from the Government, published in May 2000. In broad terms the recommendations made by the Stewart Group were accepted. It was agreed that the emissions from mobile phones and base stations should meet the international guidelines.
>
> 14. Dealing with base stations near schools, the Government agreed "schools and parents should be reassured that the base stations near schools... operate within the guidelines". The Government did not expressly

A accept the Group's recommendations concerning the area of the beam of greatest intensity, but stated that it would be working with the Group on the further issues relating to this.

> 15. There then followed, on 22nd August 2001, Planning Policy Guideline 8 (PPG 8) dealing with planning aspects of telecommunications.

[8] Since, as I shall show, it is a principal contention of the appellant
B Secretary of State that the judge misunderstood relevant provisions contained within the policy document, PPG 8, I should set out that document's material paragraphs at this stage. PPG 8 has been effective from 22 August 2001, when it replaced its predecessor of December 1992. It gives planning policy guidance in the field of telecommunications developments. The critical passages are paras 29 and 30, which are
C replicated in the same words in paras 97 and 98 of the appendix. This is what they say:

> 97. Health considerations and public concern can in principle be material considerations in determining applications for planning permission and prior approval. Whether such matters are material in a particular case is ultimately a matter for the courts. It is for the decision-maker (usually the local planning
> D authority) to determine what weight to attach to such considerations in any particular case.
>
> 98. However, it is the Government's firm view that the planning system is not the place for determining health safeguards. It remains Central Government's responsibility to decide what measures are necessary to protect public health. In the Government's view, if a proposed mobile phone base station meets the ICNIRP guidelines for public exposure it should not
> E be necessary for a local planning authority, in processing an application for planning permission or prior approval, to consider further the health aspects and concerns about them.

Those are the critical paragraphs, but I should also read para 99 of the appendix (replicating para 31 of the policy):

> F All new mobile phone base stations are expected to meet the ICNIRP guidelines. However, all applicants should include with their applications, a statement that self-certifies to the effect that the mobile phone base station when operational will meet the guidelines. In line with the Group's recommendations the mobile phone network operator should also provide to the local authority a statement for each site indicating its location, the height
> G of the antenna, the frequency and modulation characteristics, and details of power output. Where a mobile phone base station is added to an existing mast or site, the operator should confirm that the cumulative exposure will not exceed the ICNIRP guidelines.

[9] In his decision letter, the inspector noted in para 9:

> There is local concern about the proximity of 3 schools to the proposed mast,
> H and the effect of the increased output from the proposal compared to the existing mast.

In para 10, he acknowledged that the proposed installation would comply with ICNIRP guidelines, and that indeed is common ground. He then proceeded thus:

> 11. However, evidence (including material considered by Stewart) suggests that young children may be more vulnerable to the effects of electromagnetic radiation through increased thermosensitivity compared with that of the population at large. Stewart acknowledges that there is no evidence that, even under the "beam of greatest intensity", exposure to levels of electromagnetic radiation within the ICNIRP guidelines would have any harmful effect upon health. However Stewart recommends as a precautionary measure that the "beam of greatest intensity" from an installation should not fall upon schools. PPG 8 states that the Government's acceptance of the precautionary approach recommended by the Stewart Group's report "Mobile Phones and Health" is limited to the specific recommendations in the Group's report and the Government's response to them.
>
> 12. Mobile Phones and Health – the Government's Response states that "the Stewart Group recommended (1.42), in relation to macrocell base stations sited within school grounds, that the beam of greatest intensity should not fall on any part of the school grounds or buildings without agreement from the school and parents. Similar considerations should apply to macrocell base stations sited near to school grounds". The proposed installation would be sited some 200m from Woodfield Community Primary School building, and some 250m from St Roberts Catholic Primary School building; Harrogate Granby High School would be about 450m distant. Both Primary Schools would appear to have buildings and/or grounds that would be within the beam of greatest intensity. The Government Response further states: Government agrees that schools and parents should be reassured that base stations near schools… operate within guidelines. We will be working with the Stewart Group on the further issues regarding measurements of emissions from base stations on or near schools and how to take forward the recommendation on the "beam of greatest intensity".
>
> 13. The Stewart Report, the Government Response, and PPG 8 together appear to suggest that even under the "beam of greatest intensity" there would be no risk to young children from emissions within ICNIRP guideline levels. However, on the question of the recommendation on the beam of greatest intensity, the Government response is open-ended rather than conclusive. This matter is of particular relevance to the current case.
>
> 14. I conclude that the appeal proposal in its present form provides insufficient reassurance that there would be no material harm to the living conditions (in terms of health concerns) specifically of the group identified by the Stewart Report as potentially vulnerable: that is, of young children, in this case at both Woodfield Community Primary School and St Roberts Catholic Primary School.

[10] The respondents took two points upon their application under section 288: first: the inspector had misconstrued government policy, that is to say PPG 8; second: the inspector had failed to give adequate reasons for his decision. As for the first, the respondents' essential argument, advanced by Mr Christopher Katkowski QC, was that the effect, in particular of para 98 of the appendix to PPG 8, was that if international guidelines are met then that is the end of the matter. Therefore, since in this case it was beyond contention that the ICNIRP standards were met (the emissions from the equipment would be lower than those standards by a factor of thousands) the inspector should simply have been satisfied of that state of affairs.

[11] There was of course a self-certificate within the terms of para 99, and so the argument proceeded thus: the inspector had not been entitled to reject the appeal proposal on health grounds or grounds to do with the perception of health. I suppose it might be added that if he were going to take such a course, that would be a departure from the policy guidance, and there is no explanation in the decision letter of any such departure. In fact, the inspector plainly did not view his decision as departing from the policy guidance in PPG 8. He considered that it was consistent with the guidance; and that, submitted Mr Katkowski, was a misconstruction of the policy. On this first point, the judge agreed with Mr Katkowski. He said:

> 27. I prefer Mr Katkowski's submission. In my view the guidance contained in PPG 8 is perfectly clear, and there was nothing open-ended about Government policy. I have no doubt that the present proposals meet the ICNIRP guidelines for public exposure, and that it was made clear to all concerned that there would be no material harm to the living conditions (in terms of health concerns) to young children. It is also clear that the applicants gave sufficient reassurances about this.
>
> 28. Regrettably, the inspector appears to have misunderstood Government Planning Policy on this topic as set out in PPG 8 and failed to give adequate reasons for his decision.

[12] The judge's statement at the end of [28] also expresses his acceptance of the respondents' second point relating to reasons. The particular argument under that head was that the inspector had failed to give adequate reasons for his conclusion that there had been insufficient reassurance that the proposal would involve no material harm in terms of health risks.

[13] In this court, the Secretary of State submitted, first, that the judge himself had misconstrued the planning policy contained, in particular, in para 13/98 of PPG 8. It is submitted by Mr Philip Coppel that, in accepting Mr Katkowski's argument that once compliance with ICNIRP was met – and no doubt there was a certificate to that effect – that was the end of the matter, the judge implicitly concluded that once it was shown by such compliance that there was no actual risk to health, the decision-maker could not, within the policy, distinctly refuse planning permission on grounds of a perceived health risk. It is said that that is not the effect of the relevant paragraph of the policy. It is said, and certainly in this court it is common ground, that actual health risk and perceived health risk, or public concern about health, were both distinct matters treated as relevant considerations to the planning decision falling to be made, and were so treated by PPG 8.

[14] Mr Coppel's argument here may be no more than a different way of putting what I understood to be his substantive submission on para 98, which was as follows. Mr Coppel said that the judge had wrongly subordinated the statement of policy in para 97, namely that it is for the decision-maker to determine what weight to attach to considerations of health and concerns about health, to the statement of view in para 98, namely that once ICNIRP guidelines are met and certified, it should not

be necessary for the planning authority "to consider further the health A
aspects and concerns about them".

[15] It is convenient to return to the language of para 30/98. Mr Coppel submitted distinctly that where the paragraph states it should not be necessary for an authority to consider further the health aspects and concerns, the expression simply means what it says and is to be distinguished from such phraseology as "it will not be necessary so to B
do". This is an expression of view or approach, not itself a distinct policy properly so called. In addition to the language, it is submitted that the approach that Mr Coppel would advocate sits comfortably with other proposals in PPG 8, such as paras 63 and 83 of the appendix. He added that his approach accords with the general nature of planning policy guidance and makes what he calls policy sense. C

[16] This aspect of the case is at the heart of the appeal. If Mr Coppel is right then, on the face of it, the inspector was entitled to conclude (as he did) that there had been insufficient reassurance regarding public concern about health, without having to treat such an approach as being a departure from the policy. If he was wrong about it, the inspector has departed from government policy without explaining why. D

[17] The Secretary of State's next submission was that the judge had given no reasons for his conclusion that the inspector had failed to give adequate reasons for his. Then it is said that the inspector's reasons were, in any event, sufficient and intelligible; and, finally, there is a submission that the judge on an appeal on law only had not been entitled to find (as he had here) that the respondent had given insufficient reassurances E
about health concerns for young children, unless it had been perverse to find otherwise, which it was not. This clutch of arguments, in truth, adds nothing to the principal argument when the matter is examined, but I will return to them briefly.

[18] I address, first, the substantive point of construction. I do not consider that a substantial distinction can be drawn between the F
expression of policy in para 97 and the expression of a view or opinion in para 98. Paragraph 97, in truth, essentially states matters of background, the legal setting of the Secretary of State's policy, and, in para 98, the policy is expressed that if in any given case the ICNIRP guidelines are met, the planning authority should not have to look further in relation either to an actual health risk or perceived health risks. The rationale of the policy is G
the first sentence, which, to my mind, is important for an understanding of the whole. There, the Secretary of State says:

> it is the government's firm view that the planning system is not the place for determining health safeguards.

[19] What follows is drawn in the light of that first statement. It seems H
to me plain that that is as much policy as anything else in the document. Certainly, the text leaves open the possibility (and this is no more than a conventional aspect of administrative law) that there might be a case in which the planning authority would be justified in looking further and, to that extent, departing from the policy. But that would be an exceptional

course that would have to be specifically justified, as the judgment of Woolf J, as he then was, in *EC Gransden & Co Ltd* v *Secretary of State for the Environment* [1986] JPL 519, cited to the judge by Mr Katkowski, amply demonstrates.

[20] Accordingly, as it seems to me, the judge was right to hold that the inspector misunderstood PPG 8. With deference to Mr Coppel, I was not assisted by his references to other paragraphs in the appendix. It seems to me that the suggestion that the approach advocated by the Secretary of State is more in accordance with the general nature of planning policy guidance, and makes what Mr Coppel called "policy sense", is without any force. Once one recognises the thrust given to para 98 by its first sentence, this is simply a classic piece of planning policy.

[21] The inspector appears to have considered that his conclusion, that the appeal proposal provided insufficient reassurance on health, was consistent with government policy, notwithstanding the proposal's ample compliance with ICNIRP and an appropriate certificate having been given to that effect. That, in my judgment, was the error made by the inspector that is central to this case. Such a conclusion, in truth, represented a departure from the policy. Although the inspector, as I have said, might have been entitled to take such a position, he would have had to justify it as an exceptional course. I see no exceptional circumstances here, notwithstanding the fact – if it be one – that the beam of greatest intensity is directed to two of the schools. The planning policy indicated in para 98 must, in my judgment, be ample to cover such a case. In any event, the inspector did not seek to justify his conclusion by reference to anything he thought to be an exceptional circumstance. Thus, there is, as I have indicated, nothing in paras 11 to 14 to show why, on the facts of this particular case, compliance with the ICNIRP guidelines was insufficient to allay perceived fears about health issues.

[22] It is important to have in mind that this is a case in which the developers did not merely issue the self-certificate referred to in para 99; there is detailed correspondence between the developers and all three schools, as well as the local planning authority. We have been shown some of the letters this morning. I do not propose to read them out, but I will indicate that I have had particular regard to the developers' letter to the Catholic primary school of 29 January 2003, which sets out technical information. Reading the letter, I do not consider that to be a vice. The letter was written to the head teacher of the Catholic primary school. If the head teacher did not understand the letter – and I make no such suggestion – it was, of course, open to him or her to take advice about it. Moreover, the letter ends with an invitation for the school to submit representations or views that would be considered by the developers and submitted to the council. As I understand it, that was done, I think, in relation to all three schools, for there was a second round of correspondence and we have also been shown that. If one were looking to see whether the inspector had been entitled to hold that there was insufficient consultation or reassurance here, it is very difficult to my mind to see how such a conclusion could be arrived at.

[23] As far as reasons are concerned, as I have already indicated, the judge simply stated that the inspector had failed to give adequate reasons for his decision. It is said that that itself is an unreasonable conclusion. However, the reality is that the inspector had given no reasons as to why he concluded as he did. He certainly referred to such important matters as "the beam of greatest intensity" and so forth, but there is no discussion of the communications that had actually passed between the parties. In any event, we must be primarily concerned with the legal integrity of the inspector's decision letter and, in my judgment, for reasons I have given, that is fatally undermined by the inspector's misconstruction of the policy.

[24] Finally, there is the criticism of the judge's own statement that sufficient reassurances had been given. Given the view that I have taken of the construction of PPG 8, that was necessarily correct. The assurance consisted in certification of the compliance with ICNIRP, at least in the context of the letters earlier written. The judge's comment is accordingly unobjectionable.

[25] For all these reasons, in my view, the learned deputy judge was right to quash the inspector's decision, and for my part I would dismiss this appeal.

[26] **MUMMERY LJ:** I agree.

[27] **PILL LJ:** I also agree. Laws LJ has referred to PPG 8 and, in my judgment and in agreement with him, clear guidance is given there on the matters that arise in this appeal. If that guidance is to be departed from, sound reasons are to be expected. The inspector was not, as he appeared to think, making a decision in accordance with the guidance.

[28] What the inspector has sought to rely upon, notwithstanding the PPG, is a paragraph in the government's response to the report of an expert committee chaired by Sir William Stewart. The inspector has cited the government response in para 12 of his report:

> [The] Government agrees that schools and parents should be reassured that the base stations near schools... operate within guidelines. We will be working with the Stewart Group on the further issues regarding measurements of emissions from base stations on or near schools and how to take forward the recommendation on the "beam of greatest intensity".

[29] In para 13, the inspector stated that the government response is "open-ended". Reassurance was given as to the operation being within guidelines, both in the certificate supplied and in detailed explanations in correspondence, to which Laws LJ has referred. These assurances were accurate, as acknowledged by the inspector in another part of para 13 of his report. That part of the government's response, which deals with reassurances, is not open-ended. All that is open-ended in that response is the second sentence quoted by the inspector in relation to further work upon the measurements of emissions. In my judgment, the assurances required by the government response were given in this case. I agree with Laws LJ that the inspector was not following, although

A he thought he was, policy, and he had not sought to justify any departure from policy.

[30] For those reasons, and the reasons given by Laws LJ, I agree that this appeal should be dismissed.

Appeal dismissed.

R (on the application of Newsum and others) *v* Welsh Assembly Government

[2004] EWCA Civ 1565

COURT OF APPEAL
Waller, Tuckey LJJ and Sir Charles Mantell
22 November 2004

Protected species – Conservation (Natural Habitats etc) Regulations 1994 A
– Relocation of newts – Regulation 44(2)(e) – Conservation of Natural
Habitats and of Wild Fauna and Flora 92/43/EEC – Whether licence
required to relocate newts – Whether decision maker entitled to
withhold licence – Basis of granting licence to remove a protected
species – Whether preservation of species protected by regulations can
provide "an imperative reason of overriding public interest" to found B
application to relocate

The respondent trustee landowners applied to the appellant Welsh
Assembly Government for a licence, under regulation 44(2)(e) of
the Conservation (Natural Habitats etc) Regulations 1994 (the 1994
Regulations), to relocate some great crested newts from a disused quarry
to a prepared site. The respondents had planning permission to work C
the quarry. The application was refused. The respondents' application
to quash that decision was allowed by Pitchford J. The Welsh Assembly
Government appealed to the Court of Appeal, contending that, under
regulation 44(2)(e), the preservation of newts was not a matter of overriding
interest, and the question as to whether the respondents would be guilty
of an offence under regulation 40(3)(c) were they to carry out the works D
authorised by the planning permission was an immaterial consideration.
No reason of overriding public interest existed to allow the quarry to be
worked at present, and therefore no licence should be granted to relocate
the newts.

Held The appeal was allowed. E

The words "other imperative reasons of overriding public interest
including those of... beneficial consequences of primary importance to
the environment" were not intended to cover the preservation of the
species itself. Article 16 of the European Directive on the Conservation of
Natural Habitats and of Wild Fauna and Flora 92/43/EEC and regulation
44 of the 1994 Regulations are derogating provisions. All the purposes F
set out expressly in regulations 44(2)(e)-(g) are purposes that, if carried
out, will have or might have an effect upon a species or their breeding
sites. The general words "other imperative reasons of overriding public
interest" were not intended to introduce as a purpose forming the basis
for a licence, the preservation of the species. "Imperative reasons" of
"overriding public interest" contemplate a project or plan, vital in the
public interest, which will necessarily have an effect upon protected wild

A animals or wild plants: see [28]. The Welsh Assembly Government had been entitled to take the view, independent of the planning authorities, as to whether the purpose upon which the trustees relied did provide an imperative reason of overriding public interest: see [29].

Case referred to in the judgment

B *R (on the application of Newsum)* v *Welsh Assembly; sub nom Newsum* v *Welsh Assembly Government; Newsum* v *National Assembly for Wales* [2004] EWHC 50 (Admin); [2004] JPL 1537

Appeal against a decision of Pitchford J

This was an appeal a decision of Pitchford J, sitting in the Administrative
C Court on 4 February 2004, quashing a decision of the Welsh Assembly Government refusing to grant to Jeremy Newsum and fellow trustees of the estate of the Duke of Westminster a licence to move a population of great crested newts from a disused quarry to a specially prepared site. The application for a licence was made under the Conservation (Natural Habitats etc) Regulations 1994 (SI 1994/2716).

D *Milwyn Jarman QC and Andrew Keyser (instructed by the Treasury Solicitor) appeared for the appellant, the Welsh Assembly Government.*

Martin Kingston QC and David Park (instructed by Bremners) represented the respondents, Jeremy Newsum, Mark Loveday and Robin Shedden Broadhurst (trustees of the estate of the Duke of Westminster).

E *The following judgments were delivered.*

WALLER LJ:

[1] This is an appeal from the judgment of Pitchford J, dated 4 February 2004 ([2004] EWHC 50 (Admin)) by which he quashed the decision of the Welsh Assembly Government (the Assembly) refusing to grant to the trustees of the estate of the Duke of Westminster a licence to "translocate"
F a population of great crested newts (gcns) from a quarry to a specially prepared site. The application for a licence had been made under regulation 44(2)(e) of the Conservation (Natural Habitats etc) Regulations 1994[1] (the 1994 Regulations).

[2] The point that arises on the appeal, putting it very shortly for the moment, is this. Regulation 39 provides for it being an offence deliberately
G to kill gcns and (and this need not be deliberately) to damage or destroy their breeding sites. However, by regulation 40(3)(c), a person is not guilty of an offence by reason of "any act made unlawful by that regulation if he shows that the act was the incidental result of a lawful operation and could not reasonably have been avoided". Liability under regulation 39 is also excluded for anything done for certain purposes and in accordance
H with a licence obtained under regulation 44. Regulation 44(2)(e) identifies one such purpose as being "preserving public health or public safety or other imperative reasons of overriding public interest including those

[1] SI 1994/2716

of a social or economic nature and beneficial consequences of primary importance for the environment." The trustees' case is that they have planning permission that allows them to do the very things that will "incidentally" and "unavoidably" kill gcns or damage or destroy their breeding site. They say that, arguably, by virtue of regulation 40(3)(c), they will not be committing an offence if they were to do those things. They say that, in addition to the importance of allowing this quarry to be worked, it is a reason of overriding public interest that the gcns should be preserved, and a licence should be granted to relocate them. The Assembly says that, under regulation 44(2)(e), the preservation of the gcns is not a matter of overriding interest, and the question as to whether the trustees would be guilty of an offence under regulation 40(3)(c) were they to carry out the works authorised by the planning permission is an immaterial consideration. It says that there is no reason of overriding public interest in allowing the quarry to be worked at present, and thus, no licence should be granted to relocate the gcns.

Background

[3] Since there is no challenge to substantial parts of Pitchford J's judgment, it is convenient to append the same to this judgment. Both Mr Milwyn Jarman QC, for the appellant Assembly, and Mr Martin Kingston QC, for the respondent trustees, adopted as accurate parts of the judge's judgment that set out the following aspects fully and clearly: the background to the claim: in [1] to [11]; the relevant parts of the European Directive on the Conservation of Natural Habitats and of Wild Fauna and Flora 92/43/EEC (the Habitats Directive), and the 1994 Regulations: in [12] to [20]; the history of the application for a regulation 44 licence: in [21] to [27]; the correspondence leading up to the decision: in [28] to [42]; the submissions by consultees: in [43] to [49]; relevant passages of the decision letter of 14 February 2003: in [50] and [51]; and passages in the post-decision correspondence: in [52] to [53]. I do not repeat those parts but simply incorporate them as part of this judgment.

[4] In the decision letter, it is common ground that the Assembly simply concentrated on whether there was a present need for the quarrying of chert and limestone, indicating that there was, in reality, no issue that there was "currently sufficient production capacity and permitted reserves to meet local and regional demand" and that this was in line with the trustees' view, since they "had no immediate plan to reopen the quarry". The submission for a licence was (the letter records) being made to safeguard the planning permission and the value of the estate's interest in the quarries. The letter indicated that the situation might change in the future, making the site a relatively desirable source for aggregates, but stated:

> This was insufficient to demonstrate an overriding public interest now or within the timescale required to relocate the newts. If the situation in relation to overriding need for the quarry products were to change at some time in the future it would be open to [the trustees] to make a new application for a licence...

[5] The complaint of the trustees, as ultimately identified following the correspondence after the decision, was that, in the decision letter, the Assembly had: (i) ignored the existence of a valid planning permission; (ii) ignored the likelihood that in carrying out the authorised works, which would unavoidably damage the gcns, they (as the trustees contended) might be acting lawfully by virtue of regulation 40(3)(c); and (iii) had failed to have regard to the issue as to whether preservation provided a reason of overriding public interest. They further submitted that it could not be right to postpone matters until circumstances altered because it was important to take time to relocate the gcns.

[6] The Assembly's position was that the existence of the planning permission was relevant but not determinative as to whether a licence should be granted; it was not for it to decide whether, if the trustees chose to quarry and thereby damage the gcns, they would be committing an offence (but they did not accept that they would not be); and the preservation of the species that the regulations were designed to safeguard could not be "an imperative reason of overriding public interest" within regulation 44(2)(e).

[7] It was these issues to which submissions were made to the judge as recorded by him in [55] to [63], so far as Mr Kingston for the trustees is concerned, and in [64] to [75], so far as Mr Jarman for the Assembly is concerned. Those submissions were essentially repeated before us, and it is therefore not necessary to rehearse them.

[8] The key issue, albeit much background may be traversed to provide the context, is one of construction of regulation 44(2)(e), and whether the preservation of the species that the regulations are designed to protect can itself provide "an imperative reason of overriding public interest". I do not myself think that the question as to whether the trustees, if they chose to quarry, would be committing an offence is a question that needs to be answered, or indeed should be answered. It is sufficient, and no one, I think, would dispute this, that it is certainly arguable that the circumstances may be such that they might not be committing an offence, and it is certainly possible that the trustees will act in accordance with the planning permission, whether or not they obtain a licence to relocate the gcns. Thus, there is a risk that operations will take place that will kill gcns and damage or destroy their breeding site.

[9] What the trustees appear to be seeking to do is to "conserve" the gcns that are at present in the location where they wish to quarry. If they are to relocate them before they carry out any works that might incidentally damage the gcns or their breeding site, they wish deliberately to capture, and deliberately to disturb the gcns and damage or destroy their breeding site or resting place, all offences under regulation 39, unless they have a licence to do so. At first sight, one would have thought that the trustees should have been applying for a licence for the purpose of conserving the gcns under regulation 44(2)(c). That is a licence that the Countryside Council for Wales (the CCW) would issue under regulation 44(4). Before issuing, it would have to consider whether "there is a satisfactory alternative", and whether "the action authorised will be

detrimental to the maintenance of the population of the species concerned at a favourable conservation status in their range": see regulation 44(3). But the question as to whether it was necessary to start quarrying having regard to production elsewhere may not be a material consideration. The judge said, in [104]:

> Mr Jarman accepts that an application for this licence could have been appropriately addressed to the Countryside Council for Wales under regulation 44(2)(c). What then would have been material considerations for the Council? The claimants would be seeking a licence to conserve a protected species because, in light of the planning permission, it is likely the species would otherwise be harmed. If Mr Jarman's argument is sound the Council would be entitled to refuse the application simply on the ground that there was a satisfactory alternative within the meaning of regulation 44(3)(a), namely not pursuing the development under extant planning permission. In my view, that cannot be right. It seems to me the Council would be bound to proceed on the basis that the permission entitled the developer to commence operations subject to regulations 39 and 40.

[10] That is a view with which I did not understand Mr Jarman before us to quarrel. But, he did say that it is not for the Assembly to take a view of what the CCW may take into account, and it is true that this was not in fact an application to the CCW, so no submissions from it on this point have been received.

[11] I should also record that, at the commencement of the appeal, we raised the question as to whether the application should not have been made to the CCW in reliance upon regulation 44(2)(c), and Mr Jarman's submission was that it might well be that that was so, but it was not a matter for the Assembly to advise on the appropriate course for the trustees to take. Mr Kingston pointed out that, in a paragraph of a letter not quoted by the judge, Dr Christopher Betts, when writing to the Assembly, recorded that when the trustees did approach the CCW, they were told that the correct body to which to apply for a licence was the Assembly, under regulation 44(2)(e): this was the letter dated 30 September 2002, referred to in [33] of the judge's judgment. The Assembly did not seek to correct that advice and that is why matters have proceeded as they have.

[12] It was ultimately the submission of Mr Kingston that there is in fact some overlap between the regulations, and the fact that conservation is dealt with under regulation 44(2)(c) does not prevent, in a particular case, the conservation of the species from being a matter of overriding public interest with beneficial consequences of primary importance to the environment under regulation 44(2)(e). One key issue in the case is how far Mr Kingston is right about that.

[13] The judge, in his discussion of the issues, helpfully set out the planning context. The 1994 Regulations expressly deal with how planning applications, after the coming into force of the regulations, will deal with matters covered by the regulations. In essence, planning consent will not be given without a full assessment of the likely effect upon any "European site". By regulation 49, the competent authority may agree to the plan or project notwithstanding a negative assessment "if they are satisfied that,

A there being no alternative solutions, the plan or project must be carried out for imperative reasons of overriding public interest (which subject to paragraph (2) may be of social or economic nature)". Paragraph (2) provides that where the site hosts a "priority natural habitat type or priority species" the reasons must be either: "(a) reasons relating to human health, public safety or beneficial consequences of primary importance to
B the environment or (b) other reasons which in the opinion of the European Commission are imperative reasons of overriding public interest". As the judge noted, this language seems to mirror regulation 44(2)(e) and (3)(a). In this context, it is "the plan or project" that the planning authority should conclude "must be carried out for imperative reasons of overriding public interest", and that does not seem to envisage the plan or project being the
C preservation of a species.

[14] The regulations also provide for consents granted prior to the coming into force of the regulations to be reconsidered. Regulations 54-59 apply regulations 48-53 expressly to reconsideration of planning permissions. As the judge said, in [87], it follows that when permission was given to the trustees to quarry at Pen-yr-Henblas, the planning authority were not
D required to consider the effect on the site under the 1994 Regulations, but have the ability to do so pursuant to regulations 54-59. The planning authority have not however, up until this stage, taken any steps to revoke the planning permission.

[15] The judge then dealt with the proper construction of regulation 39, and whether the trustees would be committing an offence
E if they commenced quarrying and, as an incidental result, killed gcns or damaged or destroyed the breeding site of gcns. As I have said, I do not think it right to express any final view as to the width of the regulation upon which the trustees rely. The fact that it is arguable that they might not be committing an offence, and the fact that the trustees might therefore quarry and incidentally kill gcns and destroy the breeding sites of gcns
F seems to me all that it is necessary to record. I think that the judge put it correctly when he said that he was bound to approach the construction of regulation 44 with an understanding that the Habitats Directive and the regulations both recognise the possibility that a lawful operation could cause serious incidental harm to a protected species and the breeding sites thereof.

G [16] It follows that whether it is the CCW, when considering whether to grant a licence under regulation 44(2)(a)-(d), or the Assembly, when considering whether to grant a licence under regulation 44(2)(e)-(g), they too would be bound to approach their consideration on the same basis. For that reason: (a) if the CCW were considering whether to grant a licence based upon the purpose of conserving gcns under regulation 44(2)(c), it
H would not seem to be open to it simply to take the view that there was "a satisfactory alternative" of not carrying out that which the trustees had planning permission to do; and (b) if the Assembly were considering whether to grant a licence under regulations 44(2)(e)-(g), it would not be open to it simply to say a satisfactory alternative would be not to carry out that which the trustees had permission to do.

[17] I note that, in its decision letter of 14 February 2003, the Assembly A did not simply say that the alternative is not to carry out that which the trustees had permission to do, and I did not understand Mr Jarman to argue that it would be open to either regulation 44 decision maker to hold that the alternative of not doing that which the trustees had permission to do was a "satisfactory alternative" under regulation 44(3)(a), and therefore a complete answer to any application for a licence. B

[18] But, all that said, unless the preservation of the gcns could be an "imperative reason of overriding public interest" within regulation 44(2)(e), the Assembly would have no jurisdiction to grant a licence under regulation 44.

[19] The question, accordingly, is would the preservation of a species or some of a species, the subject of protection under the regulations be C an "other imperative reason of overriding public interest including those of... beneficial consequences of primary importance for the environment"? I have not found this an easy question. First, I do not find the language all that easy, but it seems that the respondents accept, and I think rightly, that we are concerned to identify an imperative reason that is both of overriding public interest and with beneficial consequences for the D environment. Second, I do not find the judge's reasoning as to why the purpose of conserving the gcns with which this case is concerned was "an imperative reason of overriding public interest" easy to follow.

[20] The judge's reasoning as to why preservation of the species could provide a reason of overriding public interest commenced with a concern as to whether, if the trustees were right, that might have the effect of the E regulation 44 decision maker being compelled "to treat protection of the species as an overriding public interest consequent upon the grant of planning permission": see [110]. He answered that anxiety with three points that developed his view as to whether preservation of the species could be an imperative reason of overriding public interest.

[21] The first point was linked with his view as to the proper construction of regulation 44(2)(c) and enabled him to refuse to construe F regulation 44(2)(e) "on the assumption... that the public interest can never be represented by a need to protect the species itself": see [113]. That may be so, but it does not follow that the licence being sought in this case was for "an *imperative* reason of *overriding* public interest."

[22] His second point was linked with his view as to whether regulation 40(3)(c) provided a defence, and also his view as to the proper G construction of regulation 44(2)(c) as well as 44(2)(e). He concluded that:

> Provided the planning authority or the regulation 44 decision maker or both addressed the admissible overriding public interest issues, then it seems to me the United Kingdom's application of the directive is indeed compliant. Whether the existence of a planning permission can form the justification for grant of a licence will depend on the circumstances of the individual case. H

But that does not point to the decision being one that must necessarily be taken under regulation 44(2)(e). Whether planning permission can form the justification for a licence will depend upon the purpose that forms the basis for the application for a licence. If the licence is sought to destroy the gcns or their breeding site, and the purpose relied upon is that quarrying will have that effect when it is carried out, then the purpose being relied upon is the quarrying and that could only be justified on the basis that the quarrying provided a reason of overriding public interest. That application could only be made under regulation 44(2)(e). In that context, it is not suggested by Mr Kingston that the simple grant of planning permission would establish that overriding public interest. If, however, the application for the licence was founded purely on the basis that the purpose was to "conserve" certain gcns by relocating them, and it was that process of relocation that would involve destroying their initial habitat and killing some gcns, then although the planning permission that might lead to the gcns and their breeding sites otherwise being destroyed could, in one sense, provide the justification, it would not be necessary that the quarrying supplied an "imperative reason of overriding public interest". It would simply be part of justifying the grant, under regulation 44(2)(c), where a reason of overriding public interest has no relevance.

[23] His third reason recognised that a regulation 44 decision maker might have to weigh up planning considerations. It was in this context that the judge concluded, in [116]:

> That judgment, it seems to me, is bound to include an assessment and whether the applicant could proceed without criminal liability under regulation 39. If so, the more weighty becomes the public interest in managing the disturbance to the advantage of the environment. Here, the Assembly was faced with an application by developers who did not need planning permission, nor was their existing planning permission subject to conditions which would prevent re-commencement of quarrying on environmental grounds or without a regulation 44 licence. In my judgment, had the Assembly correctly interpreted regulations 39, 40 and 44, it would have concluded that protection of the species was, in the circumstances of this application, of itself capable of amounting to an admissible public interest. In excluding such a public interest from its consideration and confining itself to an assessment of the demand for the limestone product, it seems to me the Assembly failed to take account of a material consideration, thus acting irrationally in the *Wednesbury* sense, and the decision should not stand.

In [116], the judge refers to "admissible public interest", but the search is for "an imperative reason of overriding public interest."

[24] The real difficulty in this case is caused by the regulations providing for two different bodies considering whether to issue licences with their respective jurisdictions defined by reference to the different purposes for which the licences were required. Article 16 of the Habitats Directive (as the judge pointed out) does not contemplate two different authorities, and in granting member states the right to "derogate" from the provisions of Article 12 "a system for the strict protection for the animal species listed..." identifies a variety of different purposes or circumstances. They could, as the judge recognised, overlap, but, if one authority was considering the matter, the fact that a purpose could fall within Article 16(1)(a), (b) or (c) etc would not matter.

[25] At one time, I considered the possibility that, on the true construction of Article 16, it was only Article 16(1)(c) that provided the

power to derogate where the purpose was to conserve a species. If that had A
been so, that would point powerfully in the direction of it being necessary to construe the words "imperative reason of overriding public interest" as including the purpose of conserving a species or even part of a species. We asked for further submissions on that thought and I accept Mr Jarman's submissions that Article 16(1)(a) would provide that power, and that it cannot be argued that the phrase "overriding public interest" *must* include B
conserving the species or part thereof.

[26] Mr Kingston's basic submission, as re-emphasised in the further submissions, is that the fact that a purpose falls within one paragraph of Article 16(1) or one sub-regulation of regulation 44(2), does not preclude it from falling within another. There is, he emphasised, an overlap and I accept that, under regulation 44(2), there are cases where the jurisdiction C
of the CCW and the Assembly might be said to overlap. An application for a licence to relocate in order to conserve could be made on the basis that public health or public safety was at risk if the relocation did not take place. Such an application could, I accept, be made under regulation 44(2)(e). It must also follow that an application to conserve that was based upon a case of imperative reason in the public interest could also be made under D
regulation 44(2)(e). But, this apparent overlap may not be a true overlap. The question would be what is "the purpose" that forms the basis of the application. If the purpose relied upon is public health, an application to the Assembly under regulation 44(2)(e) would be necessary. If the purpose is truly simply to "conserve", the application can be to the CCW under regulation 44(2)(c). E

[27] In any event, even if there is an overlap, that still does not answer the question as to whether the preservation of the species itself, or a part of the species, can amount to an "imperative reason of overriding public interest". Because I am inclined to feel that an application under regulation 44(2)(c), if made to the CCW, should succeed, there is a temptation to attempt to construe regulation 44(2)(e) in a way that would F
also enable the application to the Assembly to succeed, particularly as it seems that the trustees were advised to make their application under that regulation. However, I have ultimately concluded that it is simply impossible to construe regulation 44(2)(e) in the way suggested by Mr Kingston.

[28] I have ultimately concluded that even if there can be an overlap, G
as suggested by Mr Kingston, the words "other imperative reasons of overriding public interest including those of... beneficial consequences of primary importance to the environment" were not intended to cover the preservation of the species itself. Article 16 and regulation 44 are derogating provisions. All the purposes set out expressly in regulation 44(2)(e)-(g) are purposes that, if they are carried out, will have or might have an effect H
upon a species or their breeding sites. I cannot accept that it was likely that by the general words "other imperative reasons of overriding public interest" it was intended to introduce as a purpose forming the basis for a licence, the preservation of the species itself. "*Imperative* reasons" of "*overriding* public interest" seems to me to contemplate a project or plan,

vital in the public interest, which will necessarily have an effect upon protected wild animals or wild plants.

[29] What follows from that conclusion? If the Assembly were to have jurisdiction to be persuaded to grant a licence, that could only have been on the basis of needing a licence for a purpose other than simply conserving gcns. The only purpose that the trustees could rely upon was the existence of the planning permission and their interest in ultimately commencing their quarrying operations. As already indicated, Mr Kingston has not sought to suggest that the grant of planning permission itself demonstrates that the quarrying provides a purpose that can be described as "an imperative reason of overriding public interest." For this reason, the Assembly was entitled to take the view, independent of the planning authorities, as to whether the purpose upon which the trustees relied did provide an imperative reason of overriding public interest. On that basis, the challenge to their decision must be rejected.

[30] If I had otherwise been in favour of the respondents, I would not have concluded that there was anything in the timing point that was dealt with by the judge in [118], but, in the circumstances, there is no need to deal with that point.

[31] For the reasons that I have endeavoured to give, I would allow the appeal.

TUCKEY LJ: I agree.

SIR CHARLES MANTELL: I also agree.

Appeal allowed.

St Leger-Davey and another *v* First Secretary of State and others

[2004] EWCA Civ 1612

COURT OF APPEAL
Pill, Mummery and Laws LJJ
1 December 2004

Telecommunications development – Masts – Development control – Consideration of alternative sites – Telecommunications Act 1984 A **– Code powers in Schedule 2 – Best site – Statutory powers enabling operators to place masts on sites – Relevance of code powers to consideration of alternative sites**

Orange, a mobile communications operator, identified the roof of a police building as the preferred site for a mobile telephone mast. The police B authority would not consent to the erection of the mast on the ground that it wished to protect the roof space for operational and accommodation reasons. Orange applied for permission to erect the mast on another, less preferred location, but this was refused by the local planning authority on the grounds of the perceived risk to the health of local residents, visual intrusion, and a failure fully to consider other alternative sites. Orange's appeal to the Secretary of State was allowed by his inspector. C The inspector considered PPG 8, giving guidance on the selection of sites, and the powers that were conferred upon Orange as "operators" under para 5 of Schedule 2 to the Telecommunications Act 1984 (the code). Under the code, Orange had statutory powers to obtain a compulsory grant of a wayleave and to erect the proposed mast on an alternative site such as the roof of the police building, its preferred site, notwithstanding any lack of D consent. The inspector concluded that those powers should be used as a matter of last resort, and that therefore there was no reasonable possibility of an alternative site being available. The applications of the appellant local residents challenging that decision were dismissed by Sullivan J. In appealing his decision, the appellants contended that the inspector had misdirected himself in the application of the guidance in PPG 8 and in E relation to the availability of the powers under the code by improperly treating them as though they equated to the exercise of compulsory purchase powers.

Held The appeal was dismissed.

County courts have a limited power not to make an order on an F application under the code for the compulsory conferment of a right to erect a mast on a site in the absence of the owner's consent. The county court cannot conduct the overall assessment of the benefits and disbenefits of land use appropriate to the decision of a planning authority. The principle that a resort to the county court should not readily or routinely be contemplated was sound. Having regard to the guidance in PPG 8

A on the selection of sites, there is no duty on an applicant for planning permission to site a mast to satisfy the planning authority that it would fail on an application to the county court in respect of a meritorious alternative site: see [27]. Although planning authorities are entitled to have regard to the existence of the code powers, they are not obliged to ignore, as a material consideration, the stand taken by occupiers of sites alternative to B that advocated in the application. The inspector had been entitled to give weight to that consideration: see [28]. Neither statute nor planning policy guidance enjoins an applicant in the position of Orange to apply to the county court under the code: see [35].

Cases referred to in the judgment

Phillips v *First Secretary of State* [2003] EWHC 2415; [2003] 4 PLR 75

C *R* v *Secretary of State for Transport, ex parte de Rothschild; sub nom de Rothschild & Eranda Ltd* v *Secretary of State for Transport* [1989] 1 All ER 933; (1989) 87 LGR 511; 57 P&CR 330; [1989] 1 EGLR 19; [1989] 06 EG 123; [1988] RVR 200; [1989] JPL 173, CA

This was an appeal against a decision of Sullivan J

D This was an appeal by the appellants, Phoebe St Leger-Davey and James Harrison, against a decision of Sullivan J dismissing applications, under section 288 of the Town and Country Planning Act 1990, to quash a decision of the First Secretary of State allowing an appeal by Orange PCS Ltd against a refusal by Winchester City Council to grant planning permission for the erection of telecommunication equipment.

E *David Wolfe (instructed by Leigh Day & Co) appeared for the appellants, Phoebe St Leger-Davey and James Harrison.*

Timothy Mould (instructed by the Treasury Solicitor) represented the respondent, the First Secretary of State.

Christopher Katkowski QC and Tim Buley (instructed by Burges Salmon, F *of Bristol) represented the second interested party, Orange PCS Ltd.*

Winchester City Council did not appear and were not represented.

The following judgments were delivered.

PILL LJ:

[1] This is an appeal by Phoebe St Leger-Davey and James Harrison G (the appellants) against the judgment of Sullivan J, dated 1 March 2004, refusing an application to quash a decision of an inspector appointed by the First Secretary of State (the Secretary of State), on 19 August 2003, following a public inquiry. The inspector allowed an appeal by Orange PCS Ltd (Orange) against the refusal of Winchester City Council (the council), on 12 March 2002, to grant prior approval H for the siting and design of an 11.79m high telecommunications mast and an equipment cabin at Byron Avenue, Winchester. Orange is a telecommunications operator.

[2] The appellants live and attend school approximately 250-300m from the mast. Other issues were before the inspector, including the effect of the mast upon the character and appearance of the locality. The only issue

before the judge and this court arises out of the availability, on the evidence before the inspector, of possible alternative sites for the development.

[3] The inspector referred to the relevant development plan policy, policy TC1 in the structure plan, which provides:

> telecommunication development will be permitted provided all possible steps have been taken to minimise its environmental effects. Where there is a conflict with environmental objectives, the policy requires there to be no reasonable possibility of sharing existing facilities, no satisfactory alternative sites, and no reasonable possibility of using an existing building or other structure.

[4] Put generally, the issue is whether the inspector took account of, and correctly considered, the effect of the code power granted by the statutory telecommunications code, contained in Schedule 2 to the Telecommunications Act 1984 (the 1984 Act), since amended. Rights are conferred and obligations imposed upon operators, such as Orange, which are known as code system operators: see para 10 of PPG 8. There is a power to seek an order from the court requiring an unwilling landowner to grant the operator the right to place its equipment on the landowner's land.

[5] Planning guidance issued by the Secretary of State in PPG 8, effective from 22 August 2001, provides, in para 15:

> Local planning authorities and operators should work together to find the optimum environmental and network solution on a case-by-case basis.

Emphasis is placed, in paras 19-21 and paras 66-68 of the appendix supporting guidance, on: (i) keeping the number of masts to a minimum; (ii) the sharing of masts and sites where that represents the optimum environmental solution; and (iii) exploring the possibility of erecting antennae on an existing building, mast or other structure.

[6] Schedule 2 provides, in para 2, for agreements between operators and occupiers of land, among other things, to confer a right, for statutory purposes, to "execute any works on that land for, or in connection with the installation, maintenance, adjustment, repair or alteration of telecommunication apparatus". Paragraph 3 provides protection to occupiers of nearby land whose access may be obstructed by the works, and para 4 provides a defined right to compensation.

[7] Paragraph 5 of Schedule 2 provides:

> .– (1) Where the operator requires any person to agree for the purposes of paragraph 2 or 3 [ie obtaining the right to install equipment] above that any right should be conferred on the operator, or that any right should bind that person or any interest in land, the operator may give a notice to that person of the right and of the agreement that he requires.
>
> (2) Where the period of 28 days beginning with the giving of a notice under sub-paragraph (1) has expired without the giving of the required agreement, the operator may apply to the court for an order conferring the proposed right, or providing for it to bind any person or any interest in land, and (in either case) dispensing with the need for the agreement of the person to whom the notice was given.

> (3) The court shall make an order under this paragraph if, but only if, it is satisfied that any prejudice caused by the order –
>
> (*a*) is capable of being adequately compensated for by money; or
>
> (*b*) is outweighed by the benefit accruing from the order to the persons whose access to a telecommunication system will be secured by the order;
>
> and in determining the extent of the prejudice, and the weight of that benefit, the court shall have regard to all the circumstances and to the principle that no person should unreasonably be denied access to a telecommunication system.

Paragraph 7 empowers the court to fix financial terms where agreement has been dispensed with.

[8] These provisions have been amended and are now given effect by the Communications Act 2003, but it is common ground that the differences between the two are not material for present purposes. The expression "telecommunication system" has been replaced by a more up-to-date one.

[9] On behalf of the appellants, it is accepted that the availability of alternative sites is only exceptionally a material consideration in relation to applications for planning permission. Planning control is normally intended to restrict an owner's right to use its land as it chooses only to the extent that the public interest requires restriction on planning grounds. However, it is submitted, the existence of, and the relative merits of, possible alternative sites is, by virtue of PPG 8 and the statutory code, a material planning consideration in this context.

[10] In his decision letter, the inspector identified and described possible sites alternative to that proposed by Orange, and, in particular, a site at the headquarters of the Hampshire Constabulary and a site in the car park of Winchester railway station. He set out the health concerns expressed by the appellants and others.

[11] The inspector found, and Orange conceded, that, from a technical standpoint, the best site was the roof of a multi-storey building that forms part of the police headquarters. It was already used by two other mobile phone operators. The inspector recorded that, having considered a request to investigate the feasibility of installing the apparatus on the roof, the constabulary had stated that it was necessary to protect the roof space for its "operational and accommodation reasons". Notwithstanding efforts to persuade the constabulary to change its mind, it has said that its decision is final. The inspector stated: "Whilst this situation is regrettable, I have to conclude that there is no realistic possibility of this site being available."

[12] Although he did not quote them, the inspector plainly had in mind two letters from the constabulary. By a letter dated 25 March 2003, it stated that the "force must protect any future requirements for the roof and any additional needs for the Airwave system [a national police communications network]". A letter dated 10 April 2003 provided that: "It is necessary for the force to protect its roof space to ensure that our current accommodation review is not affected and that Airway, as a priority for emergency services, operates without interference."

[13] As to the station car park, the inspector considered the operational A possibilities but also noted that "it emerged that this site, like the police HQ, would not be available for the foreseeable future". A letter from Network Rail of 13 May 2003 had stated:

> In general, Network Rail is able to lease its land to third party telecommunications operators for the erection of their masts and antennae. However, this is only possible where stringent safety criteria are met. For B example, such masts are required to be certain distances from sensitive signalling equipment.
>
> In particular, third party masts are required to be a minimum distance of 167m from our proposed GSM-R (railway operational) masts. For this reason it is highly unlikely that Network Rail would permit the installation of a third party telecommunications system operator close to Winchester station C for the foreseeable future.
>
> I am sorry to be negative to this suggestion, but you will understand that the safe operation of the railway is of paramount importance to Network Rail and this has to take priority over the potential commercial opportunity suggested above.

[14] The inspector's conclusions were: D

> 43. The important point, in my view, is that technical compatibility is of little consequence if the landowner is unwilling to lease the land. Consequently I conclude that the station site is no different to any of the other potential alternatives where landowner consent has been withheld. It is also pertinent that the station site would not provide the required coverage of a significant proportion of the target triangle. The appellant argues that E even if the station site does become available in the future, an additional microcell installation would still be needed to provide coverage to the parts of the triangle that are best served by the appeal proposal. On the evidence of the drive-trials, there is considerable force to this argument.
>
> 44. While I have some reservations about the thoroughness of the search for alternative sites at the time the application was submitted to the Council, I am satisfied that the investigations subsequently conducted for this appeal F have been extensive and detailed. Various other streetworks options were considered by the appellant, but these would be less well camouflaged by trees and would simply transfer the perception of harm from one group of residents to another. No other realistic alternative within or close to the target area has been suggested by any party, and I saw no obvious opportunity on my visits to the area. Consequently, on the basis of the information before me, I am satisfied that there is no achievable alternative site available. G
>
> 45. In reaching this conclusion I have considered the use of the power available to code system operators under Schedule 2 paragraph 5 of the Telecommunications Act 1984. This allows an operator to compulsorily acquire land for a base station in circumstances where consent is withheld by the landowner. The appellant believes that this power has never been used by a code system operator, and states that it would not be used in this H instance, particularly against an authority such as the police or railway that provides a public service.
>
> 46. As with any form of compulsory purchase, it seems to me that this power should generally be used only as a last resort in circumstances where there is an overriding public interest and where there is no other possible course of action. I share the appellant's view that even if the use of this power

> was pursued, it would most likely be rejected by the Courts because of the existence of Byron Avenue as a feasible alternative. Furthermore, there is no mention of site acquisition by this process in PPG 8 or the associated Code of Best Practice, implying that it does not form part of the normal process of site selection. In these circumstances the provisions of Schedule 2 paragraph 5 do not alter my view that there is no reasonable possibility of an alternative site being available.

[15] Having reached that conclusion, the inspector considered in detail, by reference to the appeal site alone, what he described as the "balance of considerations". These included consideration of a direct risk to health, which he described as "extremely small", and health concerns and anxiety about the development. His conclusion was:

> 62. I have found that the proposal complies with the up-to-date policy advice in PPG 8, which seeks a balance between need and environmental impact, and with the development plan policy which is most consistent with this advice. These are matters of substantial weight, and in my judgment they significantly outweigh the very limited harm to the character and appearance of the locality and the detriment to the well-being of those living, learning and working nearby. In these circumstances, and taking into account all the other matters I have considered, I conclude that the proposal is acceptable.

Prior approval was granted, subject to conditions.

[16] For the appellants, Mr David Wolfe introduced his submissions on para 5 of Schedule 2 to the 1984 Act by outlining the development of compulsory purchase powers granted to public authorities since the mid-19th century. In the context of statutes dealing with telecommunications, there has been a trend, since the Telegraph Act 1863, to relax the circumstances in which compulsory powers may be exercised. This has culminated in the code under the 1984 Act, which provides what Mr Wolfe described as a unique and potent power. A county court is under a defined and qualified duty to make an order conferring a right to install equipment.

[17] The effect of this provision is different from that in section 34 of the 1984 Act under which the Secretary of State may authorise a public telecommunications operator to purchase land compulsorily. Such purchases are governed by the procedure in the Acquisition of Land Act 1981 (the 1981 Act) with a discretion in the Secretary of State, as confirming authority, as to whether to confirm a compulsory purchase order. That procedure, of course, applies to compulsory purchase powers conferred on many public and private authorities. The Schedule 2 power is quite different, it is submitted, in that it takes the form of a qualified duty and not merely a discretion, and the duty is upon a court and not a minister. The duty is not limited by reference to the identity of the occupier of the land.

[18] Mr Wolfe submitted that where parliament has intended to protect specific landowners against the exercise of compulsory purchase powers, it has done so expressly, as, for example, in Schedule 3 to the Electricity Act 1989, with respect to other licence holders (para 2) and land covered

by a dwellinghouse (para 6(4) of Schedule 4.) Section 45A(3) of the Civil Aviation Act 1982 is another example in protecting other licence holders. Moreover, where, unusually, the compulsory power is to be a power of last resort, that is expressly stated in the statute: see section 6(1) of the Coastal Protection Act 1949. A

[19] As a statement of the circumstances in which conventional compulsory purchase powers are exercised, reliance is placed upon the decision of this court in *R* v *Secretary of State for Transport, ex parte de Rothschild* [1989] 1 All ER 933, where powers under the Highways Act 1980 and the 1981 Act were exercised. Slade LJ, with whom Croom-Johnson LJ and Ralph Gibson LJ agreed, stated, at p938: B

> In answer to counsel's submissions as to "special rules", I summarise my conclusions thus. First, I do not accept that any special rules beyond the ordinary *Wednesbury/Ashbridge* rules fall to be applied when the court is considering a challenge to the Secretary of State's confirmation of the compulsory purchase order. Second, however, the Secretary of State, as counsel on his behalf accepted and submitted, must be satisfied that the compulsory purchase order is justified on its merits before he can properly confirm it. He must not exercise his powers capriciously. Given the obvious importance and value to land owners of their property rights, the abrogation of those rights in the exercise of his discretionary power to confirm a compulsory purchase order would, in the absence of what he perceived to be a sufficient justification on the merits, be a course which surely no reasonable Secretary of State would take. C D

[20] That is to be contrasted, submitted Mr Wolfe, with the approach to alternative sites under PPG 8 stated by Richards J in *Phillips* v *First Secretary of State* [2003] EWHC 2415 (Admin)[1], in [39]: E

> It is PPG 8 that I consider particularly important in this case. It makes consideration of alternatives an integral part of the process of assessment of an application for approval of the siting of telecommunications structures. It is true that the main thrust of the guidance with regard to alternatives concerns the *sharing* of masts and sites: applicants for new masts are expected to show that they have explored the possibility of sharing existing structures as an alternative to a new site. But, in my view, alternative new sites also fall within the scope of the guidance. The broad tenor of the guidance is to accept the principle of telecommunications structures where they are needed for coverage, but to acknowledge the sensitivity of the location of such structures and to emphasise the importance of searching, in each case, for the optimal location. The question, as it seems to me, is not just "is this an acceptable location?", but "is this the best location?", and, for the purpose of answering that question, one can, and should, look at whatever alternative possibilities there may be. F G

H

[21] In the present case, Sullivan J expressed agreement with that paragraph. His conclusions were:

[1] [2003] 4 PLR 75

> A 32. I have referred to the provisions of Schedule 2. It is plain that paragraph 5 is to be used only where agreement cannot be obtained. Moreover, in the context of operational sites, such as a police headquarters, or a railway station, the balancing exercise in subparagraph 5(3)(b) will, in substance, be little different from the exercise that would have to be carried out in evaluating the merits of a compulsory purchase order. It seems to be that it makes little difference whether one talks of an "overriding public
> B interest", or of the benefit to the public outweighing the prejudice caused to the land owner by the making of the order under paragraph 5.
>
> 33. In these circumstances, the Inspector was entitled to take the view that even if the power was exercised, the likelihood was that the court would refuse to make an order. I repeat, it was not for the Inspector to carry out a detailed exercise under paragraph 5. He simply had to decide what, on the balance of probability, was likely to be the outcome of such an application.
> C This challenge has a very narrow focus upon paragraph 5 of Schedule 2 to the 1984 Act. But the point has to be kept in proportion. This was not a public enquiry into proposal for the police headquarters site or the railway station site, it was a public inquiry into a proposal to site a mast on the Byron Avenue site. As part of that inquiry it was certainly necessary to deal in some detail with the potential of alternative sites. In effect, Dr Wolfe's submissions
> D require the Inspector to have dealt with those sites in as much detail as if they had been the subject of the appeal itself. One has to stand back and recognise the fact that although the issue of alternative sites was an important one, it was by no means the sole issue. The Inspector has dealt with the issue of alternative sites in a comprehensive manner. He specifically considered whether the suggested use of the power conferred by paragraph 5 would alter his view that there was no reasonable possibility of an alternative site
> E being available, and concluded that it would not.

[22] Mr Wolfe submitted that the question for the inspector, when considering the appellants' proposal, was: "Is this the best location?" The inspector asked that question by considering possible alternative sites, and, for example, noting the technical advantages of use of a site at the police HQ: see para 34. He did not, it is submitted, answer the question correctly
F because he excluded sites at the police HQ and the station car park merely on the basis of the respective owner's unwillingness to allow their use. He failed to express a conclusion on the sites on their comparative merits, although, it is submitted, on a fair reading of his report, the police HQ site does emerge as the best location.

[23] Having regard to the qualified duty upon the court in para 5 of
G Schedule 2 to the 1984 Act, the owner's objection to the use of a site is not a sufficient reason, it is submitted, for excluding that site from consideration. The inspector has failed to analyse, as he should have done, the nature and extent of the prejudice to the owner of the site in question. Because of the promoter's right to go to the county court, the inspector should have considered, as the court would have done, whether: (i) use of the site
H involved prejudice to the owner; (ii) any such prejudice was capable of being adequately compensated for by money; and (iii) the prejudice was outweighed by the benefit accruing to users of the system – the factors set out in para 5. The flaw in the inspector's reasoning is demonstrated, it is submitted, by his conclusion, in para 46, that "if the use of this power [to go to the county court] was pursued it would most likely be rejected by the

courts because of the existence of Byron Avenue as a feasible alternative". A
The court would have no jurisdiction to refuse an order on that ground;
its power not to make an order is limited to a consideration of the factors
set out in para 5. The court's power is not expressed, and cannot be
categorised, as a power of "last resort".

[24] I agree with one aspect of Mr Wolfe's submission, and that
is the limited scope of the power of the county court not to make an B
order if an application is made to the court. I agree with Mr Wolfe, as
did Mr Christopher Katkowski QC, for Orange, that the limitations on
the power are narrowly confined and do not permit the court to make
an environmental appraisal as between sites. The "circumstances" to
which the court, under the closing part of the paragraph, should have
regard are the circumstances relevant to determining the extent of the C
prejudice and the weight of the benefit of the site to users, as specified
in the subparagraph. It does not permit the court to conduct that overall
assessment of the benefits and disbenefits of land use that is appropriate
to the decision of a planning authority. The narrowness of the court's
power under para 3(b) of Schedule 2 is also illustrated, as Mr Katkowski
submitted, by the use, on two occasions in the subparagraph, of the word D
"access" to a system that shall not unreasonably be denied, and not to
improved access or reception.

[25] For Orange, Mr Katkowski submitted that para 15 of PPG 8 does
not impose an obligation upon applicants for planning permission to take
one of two options: either to go to the court and fail on an alternative site
or to satisfy the planning authority that they would not succeed before the E
court on an alternative site. Neither in PPG 8, nor in the *Code of Best Practice*
issued under it (Office of the Deputy Prime Minister (2002)), is there a
reference to the use of compulsory purchase powers, and the inspector
was entitled to draw attention, in planning guidance, to its absence.

[26] For the Secretary of State, Mr Timothy Mould submitted that the
paradigm case for the use of the power to go to the county court is where F
there is a single feasible site and the owner will not consent to its use. The
power can properly be categorised as a power of last resort. The respect
for property rights implicit in the inspector's finding is a legitimate one,
and the inspector's consideration as to whether alternative sites were
"achievable" was appropriate.

[27] The principle that resort to the county court should not readily G
or routinely be contemplated is, in my judgment, a sound one. The
entitlement, confirmed in Article 1 of the First Protocol to the European
Convention on Human Rights, that "every natural or legal person is
entitled to the peaceful enjoyment of his possessions" is a factor that
a planning authority may take into consideration in an application for
planning permission for the erection of a telecommunications mast on the H
land of another. I would not accept the submission (made in writing by
Mr Wolfe but disavowed orally) that an applicant must either go to the
county court and fail on a meritorious alternative site, of which there may
be more than one, or must satisfy the planning authority that it would
fail if it did so. In my judgment, there is no such duty on an applicant.

A Although a planning authority are entitled to have regard to the existence of the power in Schedule 2, they are not obliged to ignore, as a material consideration, the stand taken by occupiers of sites alternative to that advocated in the application.

[28] In this case, the inspector had been entitled to give weight to that consideration. He had before him, as a part of the written representations, B representations from responsible authorities expressing opposition to the use of their land. These included references to emergency procedures and to safety. On the basis of those representations, he had been entitled to reach the conclusion he had as to possible alternative sites.

[29] Mr Wolfe correctly submitted that the duty imposed upon the county court with regard to the right to install equipment on another's C land, together with the contents of PPG 8, puts applications for permission such as the present one in a different category from most other planning applications. Moreover, by imposing a duty on the court, the statute confers, in this limited area of "installation, maintenance, adjustment, repairs or alteration" of telecommunications systems, an approach to interference with landowners' rights different from the conventional one. D I also see force in Mr Wolfe's submission that the expression "last resort" is not a categorisation that should bind planning authorities in all cases to ignore the merits of sites because of owners' views. It is, however, a long step from that proposition to a conclusion that, on the material before him, the inspector had been required to give such weight to the existence of the power in the county court that the rights and views of landowners were E required to be overridden.

[30] In deciding what is the best location for a site, a planning authority are entitled to take those rights and views into account. The test stated by Richards J in *Phillips* (namely, is this the best location?) should not be construed so as to ignore that factor. Seeking the best location does not involve ignoring the rights and views of concerned landowners. If it is F demonstrated to the planning authority that a court is most unlikely to make an order with respect to an alternative site, that might be a powerful consideration in favour of the site proposed, but an application for permission might succeed, as this one has, without such a situation being demonstrated to the planning authority. The views of the constabulary and of Network Rail, as interested owners, could properly be taken into G account as a consideration such that the inspector had not been required in this case to make other detailed comparisons between sites or analyse how the county court would have reacted to applications.

[31] The inspector's use of the word "achievable" in his statement that "there is no achievable alternative site available" was not wholly apt in the absence of such detailed comparison and a reasoned conclusion that H an order could not have been achieved on another site. For reasons given earlier (in [24]), the court could not have declined to make a para 5 order for the reason given by the inspector: "Because of the existence of Byron Avenue as a feasible alternative." However, it was not necessary to decide whether a site could have been achieved by way of court order. The inspector had been entitled to conclude that there was no location better

than Byron Avenue on the information before him because he could take into account the views of the constabulary and Network Rail. Thus, I agree with the conclusion reached by Sullivan J in [32] and [33], although not entirely with his approval of the inspector's reasoning. A

[32] I have expressed the view that the county court's powers of refusal are, under para 5(3) of Schedule 2, limited in scope. Similarly, the existence in the county court of those powers does not require a planning authority, in every case, to conduct an analysis of how a court is likely to react to an application for an order on sites other than the site subject of the planning application. B

[33] I would dismiss this appeal.

MUMMERY LJ:

[34] I agree with the judgment of Pill LJ. C

LAWS LJ:

[35] I gratefully adopt the account of the facts and the relevant learning and statutory materials set out in the judgment of Pill LJ. I agree that this appeal should be dismissed for the reasons given by him. Neither statute nor planning policy guidance enjoins an applicant in the position of Orange to apply to the county court under para 5 of Schedule 2 to the 1984 Act (or its successor in the Communications Act 2003). Mr Wolfe's argument entails, or at least suggests, the proposition that if the county court has not been resorted to, the inspector on an appeal such as that launched by Orange should second-guess what the county court might have done. That is a position that would lack both principle and practicability. D

Appeal dismissed.

For Product Safety Concerns and Information please contact our EU representative GPSR@taylorandfrancis.com
Taylor & Francis Verlag GmbH, Kaufingerstraße 24, 80331 München, Germany

www.ingramcontent.com/pod-product-compliance
Lightning Source LLC
LaVergne TN
LVHW021112300926
849160LV00009B/701

* 9 7 8 0 7 2 8 2 0 4 6 9 0 *